中国世界级非遗文化悦读系列·寻语识遗
丛书主编 魏向清 刘润泽

中国传统木结构建筑营造技艺

（汉英对照）

郭启新 崔红叶 主编

Traditional Chinese
Timber-Frame Architecture
Craftsmanship

南京大学出版社

本书为以下项目的部分成果：

南京大学外国语学院"双一流"学科建设项目

全国科学技术名词审定委员会重点项目"中国世界级非物质文化遗产术语英译及其译名规范化建设研究"

教育部学位中心 2022 年主题案例项目"术语识遗：基于术语多模态翻译的中国非物质文化遗产对外译介与国际传播"

南京大学－江苏省人民政府外事办公室对外话语创新研究基地项目

江苏省社科基金青年项目"江苏世界级非物质文化遗产术语翻译现状与优化策略研究"（19YYC008）

江苏省社科基金青年项目"江苏世界级非遗多模态双语术语库构建研究"（23YYC008）

南京大学暑期社会实践校级特别项目"讲好中国非遗故事"校园文化活动

参与人员名单

丛书主编 魏向清　刘润泽
主　　编 郭启新　崔红叶
翻　　译 刘泽义
译　　校 Zhujun Shu　Benjamin Zwolinski
学术顾问 程小武
出版顾问 何　宁　高　方
中文审读专家（按姓氏拼音首字母排序）
　　　　　　陈　俐　丁芳芳　王笑施
英文审读专家 Colin Mackerras　Leong Liew
参编人员（按姓氏拼音首字母排序）
　　　　　　韩　煦　郝梓岑　贺　洁　梁鹏程　乔丽婷　秦　曦
　　　　　　孙文龙　吴小芳　夏　雨　杨　铮　叶继仁　叶　莹
手　　绘 陈　颖
摄　　影 陈　昊
知识图谱 王朝政
中国历史纪年简表 王朝政
特别鸣谢 江苏省非物质文化遗产保护研究所

编者前言

2019年秋天开启的这次"寻语识遗"之旅,我们师生同行,一路接力,终于抵达了第一个目的地。光阴荏苒,我们的初心、探索与坚持成为这5年奔忙的旅途中很特别,也很美好的回忆。回望这次旅程,所有的困难和克服困难的努力,如今都已经成为沿途最难忘的风景。

这期间,我们经历了前所未有的自主性文化传承的种种磨砺,创作与编译团队的坚韧与执着非同寻常。古人云,"唯其艰难,方显勇毅;唯其磨砺,始得玉成"。现在即将呈现给读者的是汉英双语对照版《中国世界级非遗文化悦读系列·寻语识遗》丛书(共10册)和中文版《中国世界级非遗文化悦读》(1册)。书中汇聚了江苏牵头申报的10项中国世界级非物质文化遗产项目内容,我们首次采用"术语"这一独特的认知线索,以对话体形式讲述中国非遗故事,更活泼生动地去诠释令我们无比自豪的中华非遗文化。

2003年,联合国教科文组织(UNESCO)第32届会议正式通过了《保护非物质文化遗产公约》(以下简称《公约》),人类非物质文化遗产保护与传承进入了全新的历史时期。20多年来,世界"文化多样性"和"人类创造力"得到前所未有的重视和保护。截至2023年12月，中国被列入《人类非物质文化遗产代表作名录》的项目数量位居世界之首（共43项），是名副其实的世界非遗大国。正如《公约》的主旨所述，非物质文化遗产是"文化多样性保护的熔炉，又是可持续发展的保证"，中国非遗文化的世界分享与国际传播将为人类文化多样性注入强大的精神动力和丰富的实践内容。事实上，我国自古就重视非物质文化遗产的保护与传承。"收百世之阙文，采千载之遗韵"，现今留存下来的卷帙浩繁的文化典籍便是记录和传承非物质文化遗产的重要载体。进入21世纪以来，中国政府以"昆曲"申遗为开端，拉开了非遗文化国际传播的大幕，中国非遗保护与传承进入国际化发展新阶段。各级政府部门、学界和业界等多方的积极努力得到了国际社会的高度认可，中国非遗文化正全面走向世界。然而，值得关注的是，虽然目前中国世界级非物质文化遗产的对外译介与国际传播实践非常活跃，但在译介理据与传播模式方面的创新意识有待加强，中国非遗文化的国际"传播力"仍有待进一步提升。

《中国世界级非遗文化悦读系列·寻语识遗》这套汉英双语丛书的编译就是我们为中国非遗文化走向世界所做的一次创新译介努力。该编译项目的缘起是南京大学翻译专业硕士教育中心特色课程"术语翻译"的教学实践与中国文化外译人才培养目标计划。我们秉持"以做促学"和"全过程培养"的教学理念，探索国别化高层次翻译专业人才培养的译者术语翻译能力提升模式，尝试走一条"教、学、研、产"相结合的翻译创新育人之路。从课堂的知识传授、学习，课后的合作研究，到翻译作品的最终产出，我们的教研创新探索结出了第一批果实。

汉英双语对照版丛书《中国世界级非遗文化悦读系列·寻语识遗》被列入江苏省"十四五"时期重点图书出版规划项目，这是对我们编译工作的莫大鼓励和鞭策。与此同时，我们受到来自国际中文教育领域多位专家顾问的启发与鼓励，又将丛书10册书的中文内容合并编成了一个合集《中国世界级非遗文化悦读》，旨在面向国际中文教育的广大师生。2023年夏天，我们这本合集的内容经教育部中外语言交流合作中心教研项目课堂试用，得到了非常积极的反馈。这使我们对将《中国世界级非遗文化悦读》用作非遗文化教材增添了信心。当然，这个中文合集版本也同样适用于国内青少年的非遗文化普及，能让他们在"悦读"过程中感受非遗文化的独特魅力。

汉英双语对照版丛书的编译理念是通过"术语"这一独特认知路径，以对话体形式编写术语故事脚本，带领读者去开启一个个"寻语识遗"的旅程。在每一段旅程中，读者可跟随故事里的主人公，循着非遗知识体系中核心术语的认知线索，去发现、去感受、去学习非遗的基本知识。这样的方式，既保留了非遗的本"真"知识，也彰显了非遗的至"善"取向，更能体现非遗的大"美"有形，是有助于深度理解中国非遗文化的一条新路。为了让读者更好地领会非遗知识之"真善美"，我们将通过二维码链接到"术语与翻译跨学科研究"公众号，计划陆续为所有的故事脚本提供汉语和英语朗读的音频，并附上由翻译硕士专业同学原创的英文短视频内容，逐步完成该丛书配套的多模态翻译传播内容。这其中更值得一提的是，我们已经为这套书配上了师生原创手绘的核心术语插图。这些非常独特的用心制作融入了当代中国青年对于中华优秀传统文化的理解与热爱。这些多模态呈现的内容与活泼的文字一起将术语承载的厚重知识内涵，以更加生动有趣的方式展现在读者面前，以更加"可爱"的方式讲好中国非遗故事。

早在10多年前，全国高校就响应北京大学发起的"非遗进校园"倡议，成立了各类非遗文化社团，并开展了很多有益的活动，初步提升了高校学生非遗文化学习的自觉意识。然而，我们发现，高校学生群体的非遗文化普及活动往往缺乏应有的知识深度，多限于一些浅层的体验性认知，远未达到文化自知的更高要求。我们所做的一项有关端午非遗文化的高校学生群体调研发现，大部分高校学生对于端午民俗的了解较为粗浅，相关非遗知识很是缺乏。试问，如果中国非遗文化不能"传下去"，又怎能"走出去"？而且，从根本上来说，没有对自身文化的充分认知，是谈不上文化自信的。"求木之长者，必固其根本；欲流之远者，必浚其泉源。"中国世界级非遗文化的对外译介与国际传播要解决的关键问题是培养国人尤其是青少年的非遗文化自知，形成真正意义上基于文化自知的文化自信，然后才有条件由内而外，加强非遗文化的对外译介与国际传播。非遗文化小书的创新编译过程正是南京大学"非遗进课堂"实践创新的成果，也是南大翻译学子学以致用、培养文化自信的过程。相信他们与老师一起探索与发现，创新与传承，译介与传播的"寻语识遗"之旅定会成为他们求学过程中一个重要的精神印记。

我们要感谢为这10个非遗项目提供专业支持的非遗研究与实践方面的专家，他们不仅给我们专业知识方面的指导和把关，而且也深深影响和激励着我们，一步一个脚印，探索出一条中国非遗文化"走出去"和"走进去"的译介之路。事实上，这次非常特别的"寻语识遗"之旅，正是因为有了越来越多的同行者而变得更加充满希望。最后，还要特别感谢南京大学外国语学院给了我们重要的出版支持，特别感谢所有参与其中的青年才俊，是他们的创意和智慧赋予了"寻语识遗"之旅始终向前的不竭动力。非遗文化悦读系列是一个开放的非遗译介实践成果系列，愿我们所开辟的这条"以译促知、以译传通"的中国非遗知识世界分享的实践之路上有越来越多的同路人，大家携手，一起为"全球文明倡议"的具体实施贡献更多的智慧与力量。

目　录
Contents

百字说明　A Brief Introduction

内容提要　Synopsis

知识图谱　Key Terms

鲁班　Lu Ban ·· 001

木结构建筑　Timber-Frame Architecture ························· 010

抬梁式　Post-and-Beam Construction ······························ 020

一屋三分　Tripartite Structure ··· 027

庑殿顶　Hip Roof ·· 037

举架　Truss Raising ·· 046

额枋　雀替　Architraves and Braces ································ 053

翼角　霸王拳　Wing-Shaped Roof Corner and Fist-Shaped
　　　　　　　Joist Head ·· 061

榫卯和斗拱　Mortise-and-Tenon Joints and *Dougong* ······· 068

藻井　Caisson Ceiling ··· 082

柱和梁　Column and Beam ·· 093

收分　Tapering ·· 105

攒尖顶	Pyramidal Roof	111
歇山顶 悬山顶 硬山顶	Gable and Hip Roof, Overhanging Gable Roof and Flushed Gable Roof	117
木牌楼	Timber Archway	133
应县木塔	Yingxian Wooden Pagoda	141
悬空寺	The Hanging Temple	157
结束语	Summary	169
中国历史纪年简表	A Brief Chronology of Chinese History	171

百字说明

中国传统木结构建筑营造技艺形成于秦汉时期，以模数制为法式，以木材为主要建筑材料，以榫卯为主要连接方法。2000多年来木结构建筑营造技艺在中国传承延绵，还远播日本和朝鲜等国家，产生重要影响。2009年，中国传统木结构建筑营造技艺入选联合国教科文组织的《人类非物质文化遗产代表作名录》。

A Brief Introduction

Traditional Chinese timber-frame architecture craftsmanship came into being during the Qin and Han dynasties. It adopts the module system as the main design rules, the wood as the primary building material, and the mortise-and-tenon joints as the dominant connection method. Over the past two thousand years, it has been inherited in China and even spread to other countries like Japan and Korea, exerting important influences on the architecture craftsmanship in China and other countries. It was inscribed on the UNESCO's Representative List of the Intangible Cultural Heritage of Humanity in 2009.

内容提要

　　小龙和大卫都对中国传统木结构建筑营造技艺有浓厚兴趣。他们不仅查阅了大量文献资料，还跟随中国古建筑专家林教授进行了实地考察。他们参观了北京故宫、应县木塔、大同悬空寺这三处中国经典木结构建筑，了解了木结构及其构件的形制、制作、功用和意义，对中国传统木结构建筑营造技艺有了更深刻的认识。

Synopsis

Xiaolong and David are interested in traditional Chinese timber-frame architecture craftsmanship. They reviewed a lot of literature and then took a field trip with Prof. Lin, an expert in ancient Chinese architecture. They visited three classic Chinese timber-frame building sites, namely, the Forbidden City in Beijing, Yingxian Wooden Pagoda, and the Hanging Temple in Datong. They gained a better understanding of traditional Chinese timber-frame architecture craftsmanship after learning the details of the form and structure of each timber component, its making function, and cultural connotation.

知识图谱
Key Terms

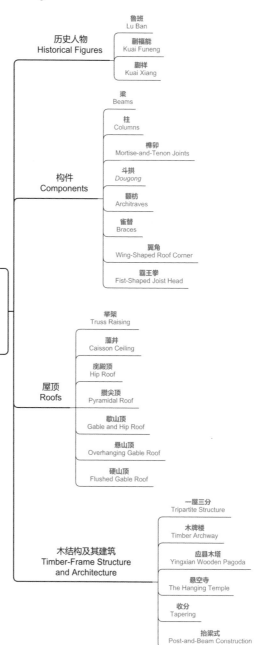

鲁班

> 小龙和大卫参加了林教授的中国传统木结构建筑营造技艺考察组。他们第一站来到举世闻名的北京故宫。

大　卫：林教授，咱们考察中国传统木结构建筑营造技艺，为什么第一站是故宫呢？

林教授：故宫有600多年历史，留下了大量保存完好的木结构古建筑，有很多可以考察的东西。

小　龙：林教授，故宫最早是谁设计修建的？

林教授：最早参与这个工程的能工巧匠很多。1405年，故宫开始修建，先由蒯（kuǎi）福能负责，后由他儿子蒯祥接替。蒯祥设计并主持修建了天安门。天安门当时称作承天门，竣工时文武百官赞不绝口，皇帝还称蒯祥"蒯鲁班"呢。

大　卫：蒯鲁班？怎么给他改名了呢？

林教授：皇帝那是夸奖他。鲁班是中国人心目中木匠的祖师爷。

大　卫：哦，我明白了。皇帝是表扬他像鲁班一样聪明。

林教授：确实是这样。他年纪轻轻就主持设计并建造这样庞大的建筑群，真得有鲁班的聪明才智才行。

小　龙：听说鲁班发明了很多木工工具。

林教授：是的。传说鲁班发明了曲尺和墨斗。曲尺可以量长短和画线。墨斗用来在木料上打直线。说起墨斗，这里面还有个故事呢。

大　卫：快讲讲，林教授，我最爱听故事了。

林教授：为了在木头上画直线，鲁班打造出前头有小洞的墨斗盒子。盒子里面有根染墨的细线，可以拽出来拉到木件顶头。拉直后弹一下细线，就会在木件上印出一条笔直的墨线。做木工活儿就有了参照。之前，鲁班的母亲每天替他拉线，很辛苦。鲁班很孝顺，不想让母亲太劳累，就动脑筋做出了这个工具。

大　卫：真巧妙啊，他怎么想到这个办法的？

林教授：他是看人钓鱼受到的启发。他照着鱼钩的样子，做了一个小铁钩，拴在线头上，把铁钩挂在木件的另

鲁班　Lu Ban

一端，这样不用别人帮忙，自己就可以弹出笔直的墨线。后来，其他木匠纷纷效仿，还把这铁钩儿叫作"替母"，因为鲁班用它代替了母亲的劳作。

大　卫：这个名字叫得好，很有趣啊。

小　龙：除了"替母"，鲁班还有其他故事吗？

林教授：有啊，还有个和故宫角楼有关的故事。据说，建造故宫四个角楼的时候，皇帝要求每个角楼都要有九梁、十八柱、七十二条脊。这可难住了工匠们。他们左试右试，就是想不出好办法。后来，来了一位神秘老人，送给他们一个蝈蝈笼子。大家发现这个笼子结构奇特，笼子里梁、柱、屋脊的数量正好符合皇帝的要求。大家顿时喜出望外。众人正要感谢这位老人家，却发现他已经消失不见了。角楼建成后，大家都说，这肯定是祖师爷鲁班显灵来帮他们的。

大　卫：看来，鲁班真是木匠心中的神呢。

林教授：是的。鲁班是个很了不起的能工巧匠。

小　龙：难怪皇帝把蒯祥叫作"蒯鲁班"呢。我们赶紧进去看看吧！

Lu Ban

> Xiaolong and David joined Prof. Lin's fieldwork trip to study traditional Chinese timber-frame architecture craftsmanship. Their first stop was the world-famous Forbidden City in Beijing.

David: Prof. Lin, why is the Forbidden City in Beijing our first stop?

Prof. Lin: The Forbidden City has a history of over 600 years. It has a lot of well-preserved ancient timber-frame buildings deserving research.

Xiaolong: Prof. Lin, who constructed it?

Prof. Lin: Many skilled craftsmen were involved in this magnificent project. Its construction started in 1405. Kuai Funeng was the chief engineer. Later, he was succeeded by his son, Kuai Xiang. Kuai Xiang designed and presided over the construction of Tian'anmen, or Chengtian Gate at that time. All of the officials were full of praise for him when it was completed, and the emperor even called him Kuai Lu Ban.

David: Kuai Lu Ban? Why was he renamed?

Prof. Lin: He was not renamed. Kuai Lu Ban was the title that the emperor used to praise him, since Lu Ban had always been considered the pioneer of carpentry in China.

David: Oh, I see. The emperor praised him for his talents.

Prof. Lin: Exactly. Since he designed and presided over the construction of massive architectural complexes at such a young age, he was very smart, just like Lu Ban.

Xiaolong: It's said that Lu Ban invented many woodworking tools.

Prof. Lin: Yes. Legend has it that Lu Ban invented the square and the ink marker for carpenters. The square is for measuring and drawing lines, and the marker for printing straight lines on timber. About the marker, there is a touching story.

David: I love stories. I can't wait for it.

Prof. Lin: To draw straight lines on timber, Lu Ban made an ink marker. Its body was composed of a small wheel, a tiny inkwell, and a long thin thread. One end of the thread was wrapped around the wheel and the other end could go through a small hole in one side of the inkwell. When this thread was pulled out onto the top of timber and flicked, a straight ink line would be printed on the timber as a reference for woodworking. At first, Lu Ban's mother toiled at pulling threads for him every day. Lu Ban was a filial son and didn't want his mother to be tired, so he made this ink marker.

David: How did he come up with the idea?

Prof. Lin: He was inspired by watching people fishing. He made a small iron hook similar to a fish hook and tied it to the thread. Fixing the iron hook to the other end of a timber, he could flick the thread and make a straight ink line all by himself. Carpenters followed his suit and called the hook "the mother substitute" since Lu Ban used it to take over his mother's work.

David: That's a brilliant name. Very interesting.

Xiaolong: Any other stories about Lu Ban?

Prof. Lin: Yes. There is one about the turrets at the four corners of the Forbidden City. It is said that during the construction, the emperor required that each turret should have 9 beams, 18 columns, and 72 ridges. The requirement was nearly impossible for the craftsmen to meet. They racked their brains but still couldn't find a way out. One day, a mysterious old man came and gave them a cricket cage. They were overjoyed when they noticed that the cage had a peculiarly perfect structure with 9 beams, 18 columns and 72 ridges to meet the emperor's requirement. But when they were about to express their thanks, the old man had disappeared. After the four turrets were built, everyone believed that the old man was a manifestation of Lu Ban, the legendary pioneer of carpentry.

David: It seems that Lu Ban has been deified by Chinese carpenters.

Prof. Lin: Yes. Lu Ban is a marvellous craftsman.

Xiaolong: No wonder the emperor praised Kuai Xiang as Kuai Lu Ban. Come on! Let's get into the Forbidden City.

木结构建筑

大　　卫：哇，好雄伟的建筑呀！

林教授：确实如此，故宫是世界文化遗产，1987年被列入联合国教科文组织的世界遗产名录。

大　　卫：故宫是中国木结构建筑的代表吗？

林教授：是的，而且故宫是世界上最大的木建筑群。

小　　龙：我知道，故宫里有9999间半房。

林教授：那只是传说。实际上大约有8700间。这里的"间"，你们知道是什么意思吗？

大　　卫：不就是一个房间吗？

林教授：这么理解不准确。中国传统木结构建筑的"间"是指"开间"，也就是四根立柱中间所形成的空间，与现代建筑的"房间"是两个概念。中国传统建筑的开间数量一般是奇数，如三、五、七、九间。开间数越多，建筑的等级越高。你们看，前面的太和殿是十一开间，而一般民居大多是三开间。

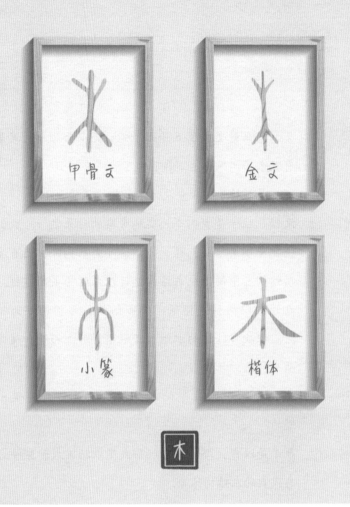

"木"字　The Chinese Character *Mu*（木）

大　卫：这么说，太和殿应该是最高等级了。

林教授：是的，这里的开间数是最多的。

小　龙：怪不得以前去参观古建筑时，人家总说这个建筑是五开间、七开间。我就很纳闷，明明是一间大屋子，怎么就变成了五开间了呢？还以为原来这里是有五个房间呢。这下我明白了。

大　卫：那就是说柱子的数量决定木结构建筑的大小？

林教授：没错。柱子是我们今天实地考察的对象之一。我们还要考察梁、枋、檩、斗拱这些大木构件，还有屋顶结构。中国传统木结构建筑以木材为主要建材，设计上以梁柱式框架、榫卯结构、斗拱构件为特点。像故宫这样的木结构宫殿还得加上彩绘和藻井两个特色。

小　龙：林教授，中国最早的木结构建筑是什么时候出现的？

林教授：最早是汉代，那时就已经形成了以抬梁式和穿斗式为代表的木结构体系。

大　卫：那些古建筑现在还能看到吗？

林教授：很可惜，看不到了。不过，古籍中有相关记载，而且有考古发现可以印证。现存最早的木结构大殿是山西五台山的南禅寺，是唐朝建筑，距今已经1200多年。

小　龙：那已经很古老了。相比于砖石建筑，木结构的保存应该更难吧？

太和殿　The Hall of Supreme Harmony

彩绘　Coloured Patterns

林教授：是的。木材易腐朽，易受虫蚁侵蚀，也易遭火灾。

大　卫：为什么不用石头呢？好像中国也不缺石料呀。欧洲建筑就会用石料。

林教授：这就是东西方文化的差异啊。我们中国人认为木出于土，向阳而生，代表着自然界旺盛的生命力。建筑为人所居，应纳天地之气，所以中国古建筑多为土木结构。

大　卫：确实跟我们的文化不一样。那故宫里用的木料应该很高级吧。

林教授：是的，故宫里用的是质地坚硬的上等木材，主要有楠木、杉木、桧木等。多数构件上还用了彩绘，既美观又实用。

大　卫：这些彩绘是很漂亮，可是有什么实际的用处呢？

林教授：可以防蛀、防潮啊。

小　龙：那就相当于油漆的作用了。这倒是一举两得。

大　卫：这么说来，这些彩绘确实既美观又实用。

Timber-Frame Architecture

David: Wow! What a majestic building!

Prof. Lin: The Forbidden City is a world cultural heritage. It was inscribed onto the World Heritage List by the UNESCO in 1987.

David: It's a masterpiece of Chinese timber-frame architecture, right?

Prof. Lin: Yes. It's the largest complex of timber-frame architecture worldwide.

Xiaolong: I know, there are 9,999 and a half rooms.

Prof. Lin: That's just a legend. Actually, there are about 8,700 rooms. I have a question for you. What does the word "room" mean here?

David: It means a room in a house, doesn't it?

Prof. Lin: Not exactly. In Chinese timber-frame architecture, the word "room" is equivalent to "space". It refers to the space defined by four columns, and we use the term *kaijian* (开间, bay) to describe it, quite different from a standard room in modern architecture. Most traditional Chinese buildings have an odd number of bays, such as a three-bay building, a five-bay building, a seven-bay building, and a nine-bay building. The more bays a building has, the higher rank it holds. The Hall of Supreme Harmony in front of us has eleven bays while most folk dwellings have only three bays.

David: Does the Hall of Supreme Harmony hold the highest rank?

Prof. Lin: Yes, eleven is the maximum.

Xiaolong: No wonder when I visited ancient buildings, phrases like five-bay buildings or seven-bay buildings were often heard. I was so confused then. I thought it meant the room had been divided into five small rooms. Now I got it.

David: So, the number of columns determines the size of timber-frame architecture?

Prof. Lin: Correct. Columns are one of the objects of our fieldwork today. Besides columns, we will also explore roofs and large timber components like beams, joists, purlins, interlocking brackets and braces. Traditional Chinese timber-frame buildings, using timber as the primary building material, feature beam-and-column frames, interlocking brackets, mortise-and-tenon joints and braces. Moreover, the large timber-frame buildings like these in the Forbidden City have two more features: coloured patterns and caisson ceilings.

Xiaolong: Prof. Lin, when did traditional Chinese timber-frame buildings appear?

Prof. Lin: By the Han Dynasty, the system of timber-frame architecture, represented by the post-and-beam construction and column-and-tie-beam construction, had already formed.

David: Can we see them today?

Prof. Lin: Unfortunately, no. But ancient documents and

archaeological findings confirm their existence. The earliest existing timber-frame hall in China was built in the Tang Dynasty, about 1,200 years ago. It's the Nanchan Temple on Mount Wutai, in Shanxi Province.

Xiaolong: That's ancient. Compared with masonry buildings, it's more difficult to preserve timber-frame ones, right?

Prof. Lin: Yes. Timber is vulnerable to decay, insect infestation, and fire.

David: Then why not build masonry buildings? It seems that there is no shortage of stone in China. Most European buildings are made of stones.

Prof. Lin: That's a cultural difference between the East and the West. We Chinese believe that wood grows from earth and thrives in the sun, so it represents the vigorous vitality of nature. And we deem it necessary for buildings to absorb the vitality of nature since they are made for people. So, most ancient Chinese buildings adopt wood and earth as the main material.

David: It's really different from our culture. The wood used in the Forbidden City must be very classy, right?

Prof. Lin: Yes, they are the finest hardwoods, including nanmu, China fir, cypress, and so on. Most timber components were painted with coloured patterns, beautiful and practical.

David: I can see these patterns are very beautiful. But what are the practical value of the coloured patterns?

Prof. Lin: They can help protect the timber from damp and worm.

Xiaolong: Just like the functions of paint. Two birds with one stone.

David: So, the coloured patterns are indeed both beautiful and practical.

抬梁式

小龙、大卫和林教授边走边聊。

小　龙：故宫的建筑都很有气势啊。

林教授：是的。故宫属于大型建筑群，采用了中国特有的抬梁式梁架结构，气势宏伟。抬梁式的房屋牢固而且耐用，还可以增大屋内使用空间，就是很耗费木材。

大　卫：抬梁式就是用柱子把梁抬起来吗？

林教授：可以这样理解。为了更准确地解释抬梁式，我们需要先说一下瓜柱和檩。

小　龙：檩我知道。中国木结构建筑是柱上有梁，梁上架檩条。檩条用来支撑椽子或屋面材料。

大　卫：那瓜柱是什么呢？也是一种柱子吗？

抬梁式　Post-and-Beam Construction

林教授：是的。瓜柱是一种立于梁、枋上面的短柱，断面可方可圆。

大　卫：就是很短很短的柱子，对吗？

林教授：是的，不过瓜柱的高度要超过它的直径。抬梁式构造是将柱子垂直立于地面，在柱子上放梁，梁上放瓜柱，瓜柱上放短梁，层层垒叠直到屋脊。最上面的顶梁上放檩，檩上放椽，椽上再放望板。

小　　龙：明白了，梁借助瓜柱一层层叠高，每一层梁逐渐缩短。

大　　卫：嗯，屋顶和梁柱是一个很稳定的三角形啊。

林教授：是的。这种结构看似简单，但要求尺寸准确。如果计算错误，木构件就组合不到一起了。

小　　龙：除了抬梁式，还有什么其他形式的构架吗？

林教授：有啊。常见的还有穿斗式和井干式。穿斗式的特点是每根柱子上直接放檩条，柱子之间用枋串接，这样可以建造大房屋。这种架构形式的柱和枋较多，虽然很牢固，但室内不能形成连通的大空间。

大　　卫：那么井干式呢？

林教授：井干式是最原始的架构形式，不用立柱和大梁，只需用圆木或矩形、六角形木料层层向上，平行堆叠，在转角处木料的端口交叉咬合，形成房屋四壁，形状如古代井上的围栏。这种房屋结构简单，但很耗费木材。

小　　龙：这么说来，穿斗式是柱网大屋，井干式是简易木屋。

林教授：是的。

Post-and-Beam Construction

Xiaolong, David, and Prof. Lin went on walking.

Xiaolong: Buildings in the Forbidden City are very imposing.

Prof. Lin: Yes. The Forbidden City is a large architectural complex with post-and-beam construction, unique and magnificent. Houses built in this way are strong and durable with larger interior space. But the disadvantage is that you have to use a lot of timber.

David: Does the post-and-beam construction mean beams are lifted by posts?

Prof. Lin: Sort of, but to fully understand the post-and-beam construction, we need to introduce dwarf columns and purlins.

Xiaolong: I know purlins. In Chinese architecture, beams are on columns, and purlins are on beams. And purlins are used to support rafters or roofing materials.

David: What are dwarf columns? Another kind of column?

Prof. Lin: Yes. The dwarf column is a kind of short column standing on the tops of beams and joists. Their cross sections can be circular or rectangular.

David: These columns are very short, aren't they?

Prof. Lin: Yes, but the height of dwarf columns should be greater than their diametres. Now, let me explain the post-and-beam construction in detail. First, columns are positioned perpendicular to the ground and beams are placed on columns. Then, dwarf columns are laid on the beams, and shorter beams on the dwarf columns. Back and forth, it goes until the multiple layers reach the ridge. At last, purlins are placed on the last layer of the short beams, rafters on purlins, and roof boarding on rafters.

Xiaolong: Oh, I see. With beams lifted higher and higher by dwarf columns, the length of every layer of beams is gradually shortened upwards.

David: Wow, it's a very strong triangular structure.

Prof. Lin: Exactly. This construction seems simple but requires accurate calculation. Timber components won't fit each other with any miscalculation.

Xiaolong: Are there other kinds of construction?

Prof. Lin: Of course. The column-and-tie-beam construction and the log-cabin construction are another two common styles. The column-and-tie-beam construction can be used to build large buildings as well. In this style, purlins are placed directly on columns which are penetrated by joists for good connection. However, with so many columns and joists, it's impossible to have a large space.

David: What about the log-cabin construction?

Prof. Lin: It's a primitive form of architecture. Round, rectangular or hexagonal timber is used to stack parallel layers without any column and beam. Corners of these timbers are intersected with each other, forming four walls of the building, just like the shape of the wooden fence of ancient wells. It's simple in construction, but still uses a lot of timber.

Xiaolong: I see. The column-and-tie-beam construction makes large houses with networks of columns, while the log-cabin construction makes simple wooden houses.

Prof. Lin: Exactly.

一屋三分

> 小龙、大卫和林教授站在太和殿前八米多高的须弥座高台上。大卫注意到汉白玉护栏上的雕刻。

大　卫：小龙，你看，这些雕刻太美了！好像雕刻的是鸟。

林教授：大卫，那可不是普通的鸟，是凤凰。这些汉白玉栏板和望柱上雕的是龙凤纹。我们现在站的地方叫须弥座高台。

小　龙：什么是须弥座？

林教授：须弥座本来是安放佛像的台座，后来用在建筑设计上，主要是为了凸显这处建筑尊贵的地位。你们看，故宫中轴线上的太和殿、中和殿和保和殿三大殿都建在须弥座高台上。

大　卫：须弥座好大呀！那栏杆下面伸出来的是龙头吧？

林教授：是的。这些龙头不仅能显示皇家威严，而且还很

门海　Bronze Water Vat

实用。

大　卫：是让栏杆更结实吗？

林教授：不，是用来排水的。三大殿高台上的龙头有1100多个，下雨天就会出现"千龙吐水"的奇观。

大　卫：那一定非常壮观。林教授，我还有一个问题，整个故宫都是木建筑，怎么防火呢？

林教授：问得好。你们来看太和殿的两侧有什么？

小　龙：那不是大水缸吗？个头可真够大呀。

林教授：这就是故宫过去的消防工具——门海，每个能装3吨水呢。为了防火，各个宫殿外面都有门海。我们刚才从午门进来，不是过了一个桥吗？桥下就是故宫的内金水河，那是救火取水的地方。

小　龙：看来古人不仅有消防意识，还很有办法呢。

林教授：没错。中国木结构建筑体现了很多传统智慧。你们听说过"一屋三分"吗？

大　卫：一屋三分？是说分左、中、右三部分吗？

林教授：不是的。是指屋体从下往上划分成台基、屋身、屋顶三个部分。

小　龙：林教授，咱们脚下的高台就是台基吗？

林教授：这还不是台基。台基指的是建筑的底座，房屋的台

一屋三分 Tripartite Structure

基有防水、避潮的作用。

小　龙：我明白了。咱们脚下的是须弥高台，屋檐下面那个台子才是台基。

林教授：对。台基是在屋檐滴水的范围之内。你们看，屋檐最前端像一片连绵的波浪线，上面圆形的是瓦当，朝下的三角形部件叫滴水。

大　卫：屋身就是台基和屋顶之间的那部分吧？

林教授：是的。屋身由木框架和墙壁构成。有一句俗语叫"墙倒房不塌"，这是说墙壁只起辅助作用。梁、柱形成的木框架才是房屋的主要支撑结构。

小　龙：也就是说，屋顶的承重主要是靠梁和柱，对吧？

林教授：是的。当然，在传统木建筑中，屋顶非常重要，也最有特色。屋顶在高度、形制等方面有严格的等级区分。太和殿是故宫内体量最大的宫殿，屋顶的高度最高，级别也最高。我们今天实地考察的重点之一就是屋顶。

小　龙：看来屋顶设计里面大有学问呢。

林教授：是的。平整开阔的台基、稳如泰山的屋身、造型别致的屋顶，共同成就了中国木结构建筑之美。

Tripartite Structure

> Xiaolong, David, and Prof. Lin were standing on the sumeru pedestal, higher than eight metres. David was attracted by the carvings on the white marble guardrails.

David: Xiaolong, look! How beautiful these carvings are! They seem to be birds.

Prof. Lin: David, they are actually phoenixes, a kind of legendary birds in Chinese mythology. The carvings on the white marble protective boards and guardrails are Chinese dragons and phoenixes. Besides, the platform we are now standing on is a sumeru pedestal.

Xiaolong: Sumeru pedestal? What's that?

Prof. Lin: Originally, it was for supporting the statues of Buddha and later adopted in architecture to highlight the dignity of the building on it. Look! The three main halls of the Forbidden City, that is, the Hall of Supreme Harmony, the Hall of Central Harmony and the Hall of Preserving Harmony, stand all on the sumeru pedestal.

David: How grand the sumeru pedestal is! And these protruding heads are dragon heads, right?

Prof. Lin: Yes. These Chinese dragon heads symbolise the royal authority and they are very practical as well.

David: To make handrails sturdier?

Prof. Lin: No, they are for drainage. There are more than 1,100 dragon heads on the platforms, making the wonder "thousands of dragons spouting water" on rainy days.

David: It must be very spectacular. Prof. Lin, I've one more question. All buildings in the Forbidden City are timber-framed, so how did they prevent fire?

Prof. Lin: Good question! Look! What are the objects on each side of the Hall of Supreme Harmony?

Xiaolong: Big water vats? How huge they are!

Prof. Lin: They used to be the firefighting equipment in the

Forbidden City, named figuratively "Sea at the Gate" in Chinese. Each bronze water vat can store three tons of water. They were installed outside every hall to prevent fire. Do you remember the bridge we crossed shortly after we went through the Meridian Gate? Under that bridge is the Inner Golden Water River, the water source.

Xiaolong: It seems that our ancestors were not only aware of the fire threat, but also good at taking precautions.

Prof. Lin: Indeed. These timber-frame buildings reflect the wisdom of our ancestors. By the way, have you ever heard of the term "tripartite structure"?

David: Tripartite structure? Does it mean a building is divided into three parts: the left, the central, and the right?

Prof. Lin: Not so. It means a building is vertically divided into three parts, namely, the foundation, the body and the roof.

Xiaolong: Prof. Lin, aren't we standing on a foundation now?

Prof. Lin: No. A foundation refers to the base of a building. Its function is to resist water and damp.

Xiaolong: I see. The platform we are standing on now is a sumeru pedestal and that platform under the eaves is a foundation.

Prof. Lin: Yes. The foundation is a platform within the area below the eaves of buildings. Look! The front end of the eaves is like a continuous wavy line, with the upper round parts called cap tiles and the lower triangular parts dripping tiles.

David: And the body refers to the part between the foundation and the roof, right?

Prof. Lin: Right. It's composed of timber frames and walls. About their functions, there is an old saying, "A timber-frame building won't collapse even if its walls fall down." It means that walls only play a minor role in supporting the building, and the timber frame with beams and columns is the main load-bearing structure.

David: That is to say, the roof is mainly supported by beams and columns.

Prof. Lin: Exactly. As for the roof, it is a very important

and distinctive part in traditional timber-frame architecture. Its rank is judged through its height and shape. The Hall of Supreme Harmony is the largest palace in the Forbidden City. Its roof is the highest, and thus it holds the highest rank. Today, another focus of our study is the roof.

Xiaolong: There is quite a lot to learn about roof designing, I think.

Prof. Lin: Yes. The flat foundation, the steady body, and the unique roof all contribute to the beauty of ancient Chinese architecture.

庑殿顶

小龙、大卫和林教授在须弥台上仰望着太和殿。

大　卫：这个殿可真大呀。

林教授：是啊，这是故宫中最雄伟的大殿，也是中国现存最大的木结构大殿。它宽约64米，深约37米，高约27米。

小　龙：这个屋顶很特别，有四面坡呢。

林教授：这叫庑殿顶。前后屋面相交，形成一条正脊，两侧屋面与前后屋面相交，形成四条垂脊，因此又叫五脊四坡式。

大　卫：看，这屋檐有两层呢。

林教授：是的，这是重檐庑殿顶。庑殿顶只能用于皇家、宗教场所等高等级建筑。庑殿顶再增加一层屋檐，就是重檐庑殿顶，是等级最高的屋顶。

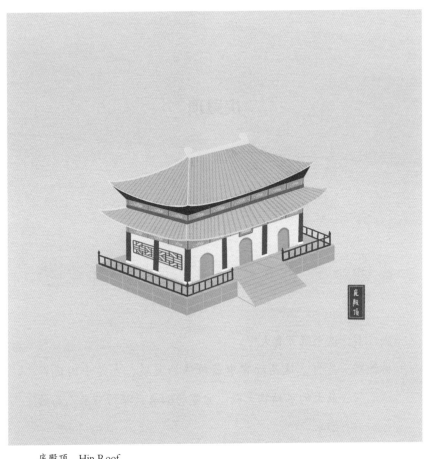

庑殿顶　　Hip Roof

小　龙：哦，看来中国古代建筑形制非常严格。

林教授：是的，过去的建筑形制等级森严，绝对不能越级。太和殿是故宫里等级最高的建筑，它不只是一座宫殿，也是最高统治的象征。这里是举办各种重要典礼的地方，如皇帝登基、点将出征等。每年万寿

庑殿顶　Hip Roof　039

太和殿　The Hall of Supreme Harmony

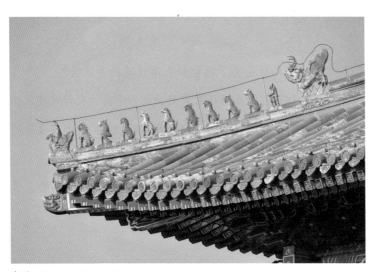

脊兽　Anti-Evil Tile Nails

节、元旦、冬至三大节①,皇帝都会在此接受文武百官朝贺,赐宴招待王公大臣。

小　　龙：看来这些古建筑也是中国古代历史的见证呢。

林教授：确实如此。

大　　卫：林教授,您看,屋脊上面有一个小人和10个小动物的雕塑,那是什么呀?

林教授：那叫瓦镇,那个小人是骑凤仙人,10个小动物都是神兽,叫镇瓦兽。

大　　卫：瓦镇？镇瓦兽？都是什么意思呢？

林教授：瓦镇的意思是压瓦镇邪,是屋顶上的镇邪物。镇瓦兽的数量越多,说明建筑物等级越高,数量最多是11个。

小　　龙：那它们就是为了镇邪和表示建筑物等级吗？

林教授：不仅如此,瓦镇还有实用功能。中国古代传统建筑正脊和垂脊衔接的地方都会有瓦镇,它们有固定屋脊的功能。

大　　卫：这些瓦镇和屋顶上的瓦都是黄色的,金灿灿的,很好看。

林教授：是的。这种瓦叫琉璃瓦，是一种高级彩色屋面材料，不仅美观，而且结实耐用。黄色琉璃瓦仅限皇家专用。

小　龙：那就是瓦的顶级配置了。

注释：

① 万寿节、元旦和冬至：清朝三大节日。万寿节是皇帝的生日。元旦指的是阴历大年初一。冬至兼具自然与人文两方面的意义，既是二十四节气中一个重要的节气，也是中国民间的传统祭祖日。

Hip Roof

> Standing on the sumeru pedestal, Xiaolong, David, and Prof. Lin looked up at the Hall of Supreme Harmony.

Xiaolong: How grand!

Prof. Lin: Yes. It's the grandest hall in the Forbidden City and the largest existing timber-frame hall in China, about 64 metres wide, 37 metres deep, and 27 metres high.

David: This roof is very special. It has four slopes.

Prof. Lin: It's a hip roof. It has a principal ridge on the intersection of the front and back slopes and four diagonal ridges on the intersections of the four slopes. Hence, it is also called "a roof with five ridges and four slopes".

David: It seems that there are two layers of eaves.

Prof. Lin: Yes. It is a hip roof with double eaves. In the past, the hip roofs were only used in the high-ranking buildings, such as royal and religious ones. The hip roof with two eaves signifies the highest rank.

Xiaolong: It seems that the architecture regulations were really strict in ancient China.

Prof. Lin: Indeed, they were compulsory and nobody could ignore them. The Hall of Supreme Harmony, the top-ranking building in the Forbidden City, was not only a palace but also a symbol of supreme power. Many important ceremonies, such as emperors' enthronements and the start-out of soldiers' expeditions, were held here. Besides, the emperors would summon civil and military officials, accept their greetings and hold banquets in this hall on three major festivals of the Qing Dynasty[1] — the Emperor's Birthday, New Year's Day, and Winter Solstice Festival.

Xiaolong: Wow, these ancient buildings literally witnessed the Chinese history.

Prof. Lin: Exactly.

David: Prof. Lin, look! There's a small figure statue and ten animal sculptures on the top of the ridge. What are they?

Prof. Lin: They're anti-evil tile nails. The figure is an immortal riding a phoenix and the ten sculptures are the auspicious animals used to decorate tile nails, called anti-evil zoomorphic tile nails.

David: Anti-evil zoomorphic tile nails? What do they mean?

Prof. Lin: The anti-evil tile nails are used to fix roof tiles and believed to be able to protect the building from evil things. The more auspicious animals a building has, the higher rank it holds. And the upper limit for auspicious animals is eleven.

Xiaolong: So, they're used for protecting the building from evil things and for manifesting the architectural hierarchy?

Prof. Lin: More than that. Their basic practical function is to fix ridges, so they're very common in ancient Chinese architecture.

David: Both the anti-evil tile nails and the tiles are yellow, glittering like gold. Very beautiful.

Prof. Lin: Yes. They are glazed tiles, a high-grade coloured roof material. It is not only beautiful, but also durable. By the way, the yellow glazed tiles are used exclusively by royalty.

Xiaolong: I got it. They're first-class.

Note:

1. **Three Major Festivals of the Qing Dynasty**: The Emperor's Birthday, New Year's Day, and Winter Solstice Festival. The Emperor's Birthday was an annual public holiday celebrating the birthday of the reigning emperor. New Year's Day is the first day in traditional Chinese calendar. Winter Solstice Festival has both natural and humanistic meanings: it is significant among the 24 solar terms, and it is also a traditional festival for Chinese to offer sacrifices to their ancestors.

举架

> 小龙注意到了太和殿垂脊优美的曲线。

小　　龙：林教授，从侧面看,太和殿的垂脊好像不是直的,而是一个凹形曲线。

林教授：嗯,没错,那是因为太和殿的屋顶是一个曲面。说起中国古建筑,最富有特色的地方就是屋脊的曲线设计,你们再仔细看看。

小　　龙：是啊,屋脊凹曲,屋檐远远伸出,屋角微微翘起,给人一种非常灵动的感觉。

大　　卫：对,是特别好看。

林教授：屋脊的这种曲线设计不仅造型优美,还有利于采光和排水。

大　　卫：以前的人是怎么设计这种曲面的呢？

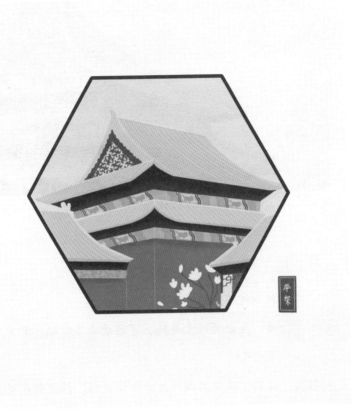

举架 Truss Raising

林教授：这就要说到举架了。举架是一种传统的木结构建筑营造技艺，在古建筑的屋顶设计中应用广泛。

大　卫：举架是什么意思呢？

林教授：可以理解成将屋架逐渐举高，以调节屋面的高度，

太和殿垂脊　Diagonal Ridges for the Hip Roofs on the Hall of Supreme Harmony

只不过举高的幅度不是平均的。

小　　龙：噢，屋面越往上越陡，那每次举高的幅度肯定是越来越大吧？

林教授：没错。这里用到的关键部件是檩。檩是架在梁上的水平构件，与屋顶的正脊平行。檩组合起来可以形成托起屋顶的架构。最上面的是脊檩。两侧檩的数量相等，相邻两个檩之间的水平距离都是一样的。

小　　龙：那就是说，檩和檩之间的垂直距离会有变化，这样屋顶才会有曲面坡度变化，对吧？

林教授：对。用举架这个技艺，檩和檩之间的垂直距离要参照各檩之间的水平距离来确定。

小　龙：明白了，横向距离相等，垂直距离不等，而且自下而上逐渐增加，这样也就形成了越来越陡的屋顶曲面。

林教授：理解得很对。

Truss Raising

> Xiaolong was attracted by the beautiful curves of the diagonal ridges on the Hall of Supreme Harmony.

Xiaolong: Prof. Lin, it seems that these diagonal ridges are not straight, but slightly concave.

Prof. Lin: You're right. The roof of the Hall of Supreme Harmony is curved. Actually, the most distinctive feature of ancient Chinese architecture lies in these curved ridges. Take a closer look.

Xiaolong: Concave ridges, extending eaves, and slightly upturned roof corners. They're beautifully designed.

David: Yes, and very attractive.

Prof. Lin: This design is not only beautiful but also practical. It's helpful for daylight and drainage.

Xiaolong: How did the ancients come up with this design?

Prof. Lin: Well, it's a traditional timber-frame architecture craftsmanship called *jujia* (举架), or the method of truss-raising. It was widely used in roof design of ancient buildings.

David: What does the term mean?

Prof. Lin: Simply speaking, it means raising the truss gradually. And the height of each "raising" is different.

Xiaolong: Oh, I see. The roof gets steeper and steeper upwards, and the height of each "raising" must be increasing.

Prof. Lin: Exactly. This method involves a key component — purlin. It's a horizontal component on the beam, parallel to the principal ridge. Purlins are used to support roofs. The uppermost one is the ridge purlin, and others are distributed evenly on the two sides. The horizontal distance between two contiguous purlins is the same.

Xiaolong: And the vertical distance between two contiguous purlins is different. Then the roof will be a concave curved surface, right?

Prof. Lin: Correct. The vertical distance between two contiguous purlins is calculated based on their horizontal distance in this method.

Xiaolong: I see. The horizontal distances are all equal, and the vertical distances are not. With the vertical distance increasing from the bottom purlin to the top one, the roof becomes steeper and steeper. Hence, a concave roof is formed.

Prof. Lin: Yes.

额枋　雀替

大　卫：林教授，这个横木上画的是二龙戏珠吧，中间的龙还有点儿立体感呢。

林教授：说的没错，是二龙戏珠。你们知道这个画着二龙戏珠的横木叫什么吗？

小　龙：我在介绍故宫的书上看到过，让我想一下……嗯，

二龙戏珠额枋　The Architrave with a Painting of "Two Dragons Playing with a Pearl"

大小额枋　Large and Small Architraves

　　　　　是额枋吧？

林教授：是的，额枋是枋的一种，也叫檐枋。枋是木建筑构件的一大类，可以细分成11种。按照枋的位置，可以分为额枋、金枋、脊枋等。枋的位置不同，作用不同。有的枋受力，有的枋起连接作用，有的既受力也能连接。

大　卫：这里上下两个都是额枋吗？

林教授：是的。你们仔细看，这里额枋可不止两块。庑殿顶下面那块叫上额枋，重檐下的两块，上面的叫大额枋，下面的叫小额枋，它们连接着相邻的两个檐柱。你们看到上额枋和大额枋上面各有一块一通到底的窄板了吗？

大　卫：就是在柱子顶上那个细长的窄板吗？

林教授：是的，那叫平板枋。这些枋都承受着来自屋顶的重量。

小　龙：那就是说枋也是承重构件了？

林教授：枋类不是主要的承重构件，主要起横向联系作用。等进到殿里面，梁架和枋之间的关系会看得更清楚些。

大　卫：小额枋下面的两个小翅膀雕刻得真漂亮呀。

林教授：那个叫雀替，也是传统木结构建筑中常见的构件。

大　卫：为什么叫雀替呀？

林教授：雀替这名字背后还有个故事呢。传说，古时有位独居老妇人曾为一只云雀疗伤，云雀伤好后就一直陪着她。一天晚上，狂风肆虐，暴雨如注，眼看老妇人的房子就要倒塌。这时，云雀飞到梁柱之间，撑开双翅，变成了一个木构件，撑住了房屋。妇人转危为安，心怀感激，就把这个木构件叫作雀替。

小　龙：这么说，雀替是起加固作用的。

林教授：是的。雀替是一种辅助性构件，安置在梁或额枋与

额枋　Architraves　　雀替　Braces

柱的交接处，承托梁枋。它可以缩短梁或枋之间的净跨度，提高承载力。不过，雀替也在不断演变，从最初的承重构件，逐渐演变成一种纯装饰性构件。雀替外形经常雕刻成各种具有美好寓意的样式，不光是鸟雀，还有龙凤、花草等纹样。

Architraves and Braces

David: Prof. Lin, is the painting on this horizontal component called "two dragons playing with a pearl"? The dragon pattern in the middle looks vivid.

Prof. Lin: Yes, indeed. Do you know the term for this horizontal plank?

Xiaolong: I once saw it in a book introducing the Forbidden City. Well, let me see... Is it called an architrave?

Prof. Lin: Yes. It's a kind of joist, called the eave joist. The joist is a major category of timber-frame architecture components. It can be subdivided into 11 categories in total. They are architraves, golden joists and ridge joists according to their positions. They have different functions. Some are used for load-bearing; some for connection; and still others combine the two functions.

David: Are the two planks here both architraves?

Prof. Lin: Yes. Look! Actually, there're more than two. The wide one under the hip roof is called the upper architrave. Under the lower eaves, the upper one is called large architrave and the lower one, the small architrave. These two architraves are used to connect the two adjacent eave columns. Have you noticed the two narrow planks above the upper architrave and the large architrave?

David: Do you mean the narrow planks above the columns?

Prof. Lin: Yes. They are plate joists. All of these joists are used to bear the weight of the roof.

Xiaolong: So the architraves are also load-bearing components?

Prof. Lin: Yes, they're, but bearing weight is not their main function. Their main function is to horizontally connect columns. We can see the difference between beams and joists clearly when we get into the hall.

David: Prof. Lin, the two small wings below the small architrave are very beautiful.

Prof. Lin: It's a pair of braces, a common component of traditional timber-frame architecture. There's a story about this name.

David: What's that?

Prof. Lin: Well, legend has it that once upon a time, there was an elderly woman who lived alone. She once healed a skylark, and the bird stayed with her ever since. One night, the wind was raging, and rain pouring. Her house was about to collapse. Suddenly, the skylark flew to the place where beams and columns intersected, spread its wings and turned itself into a wood component which helped to prop up the house. The woman survived in the end. She was very grateful to the bird, and thus called the component *queti* (雀替), meaning literally "skylark substitute". Ever after, it's called brace technically.

Xiaolong: So, this component is a reinforcement.

Prof. Lin: Yes. It's an auxiliary component placed at the junction of beams or that of architraves and columns to help support beams and joists. It shortens the span between beams or joists and improves the load-bearing capacity. The brace, though small, has evolved constantly over time, from a load-bearing component to a solely decorative one. It's often carved into various shapes with auspicious connotations, such as birds, Chinese dragons, phoenixes, flowers, and plants.

翼角　霸王拳

林教授：现在再考考你们，你们知道脊兽下面的屋檐转角部分叫什么吗？

大　卫：这个？我不知道。

林教授：这个结构叫翼角。翼角上翘是一种特殊的屋角做法，始于南北朝后期。

小　龙：哦，这个形状有点像鸟的翅膀，所以叫翼角。

大　卫：可这里很平，我看不像鸟的翅膀。

林教授：大卫观察得很仔细。这个翼角确实比较平缓。这是北方建筑的特点，南方建筑的翼角就更像鸟翼一些。

大　卫：您这么一说，我就想起来了，苏州园林里的房子和亭子都有翘翘的翼角，很好看。

小　龙：南北方翼角的形态为什么不一样呢？

林教授：这和地域气候有关。北方冬季多雪，翼角起翘幅度不

翼角　Wing-Shaped Roof Corner

翼角　Wing-Shaped Roof Corner

大，比较平坦，可减少积雪对末端瓦片的压力。南方雨水多，常有暴雨，翘起的翼角有利于排水和通风。北方的翼角看上去更沉稳大气，南方的翼角更轻巧灵动，各有各的美。

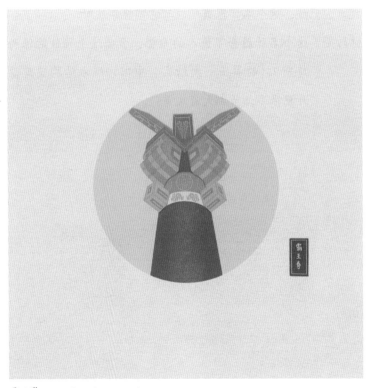

霸王拳　Fist-Shaped Joist Head

小　　龙：明白了，建筑设计首先要考虑实用性，然后才是美感。

大　　卫：林教授，您看，这个柱子上凸出来的木头也挺美的，上面还有画呢。

林教授：这个构件起到稳固角柱和额枋的作用，通常处理成凸凹型，上有彩绘，叫霸王拳。

大　　卫：哇，这个名字很霸气。可它不像拳头呀。

林教授：这个名称蕴含了匠人的希望，希望这个构件能像西楚霸王①的拳头一般结实。中国传统木结构建筑设计里很多细节都大有文章。

注释：

① 西楚霸王：即项羽（公元前232年—公元前202年），秦朝末年与刘邦争天下的楚国贵族，自立为西楚霸王，建立西楚政权，定都彭城（今江苏徐州），以孔武有力而闻名。

Wing-Shaped Roof Corner and Fist-Shaped Joist Head

Prof. Lin: A question for you guys. Do you know the term for the roof corner under these figurines of Chinese auspicious animals?

David: That? I'm not sure…

Prof. Lin: It is a wing-shaped roof corner. It's a unique construction of the roof corners, originating from the Northern and Southern dynasties.

Xiaolong: Oh, it looks like the wings of birds. That's the reason for its name, right?

David: But the corner looks quite flat here.

Prof. Lin: It was very sharp of you to see that, David. This one is indeed flat, typical in northern China. But in southern China, they are much more like bird wings.

David: Southern China? That reminds me of houses and

pavilions in Suzhou Gardens. Their roof corners are bending upwards, quite beautiful.

Xiaolong: Why are the roof corners different in the north and the south?

Prof. Lin: They are different because of regional climates. In northern China where it often snows, they are made flat to reduce the snow load on tile ends. In southern China, however, it is rainy, so they are clearly curved, very helpful for drainage and ventilation. The roof corners in the north generate a feeling of stability and dignity, while those in the south give a feeling of lightness and flexibility. Each has its own beauty.

Xiaolong: I see. In architecture, usefulness always outweighs aesthetics.

David: Prof. Lin, look! This protruding part on the top of the column is also beautiful. There're attractive paintings on it.

Prof. Lin: This part is to stabilise the corner columns and architraves. It's usually made into a concavo-convex shape with paintings. It's called a fist-shaped joist head.

Xiaolong: Wow, what a powerful name! But why is it called "fist-shaped"? It doesn't look like a fist.

Prof. Lin: It bears the hope that the intersection part would be as firm as the fists of the Overlord of Western Chu[1]. There is a lot of cultural information behind all these subtle parts in traditional Chinese timber-frame buildings.

Note:

1. **The Overlord of Western Chu:** Xiang Yu (232 B.C. – 202 B.C.), a noble of the Chu State who rebelled against the Qin Dynasty and battled with Liu Bang to be the emperor. He enthroned himself as the "Overlord of Western Chu" and founded the Western Chu Regime, in Pengcheng, now called Xuzhou, in Jiangsu Province. He is widely known for his physical strength and courage.

榫卯和斗拱

> 小龙、大卫和林教授来到太和殿入口处。

林教授：小龙,大卫,我们去看看大殿的木结构。你们看,额枋上面这些构件很精巧吧。

小　龙：您说的是这些斗拱吗?我知道这是中国建筑独有的构件。

林教授：没错,斗拱是中国传统建筑特有的构件,春秋时期就出现了,历史悠久。斗拱通常用来提升建筑整体结构的稳定性,也具有观赏性。太和殿的斗拱特别精巧,就很有代表性。但要了解斗拱,需要先知道它的结构。

小　龙：是榫卯结构吧。书上说榫卯结构不用钉子。不过我还是不太明白,不用钉子怎么能行呢?

林教授：那得先从传统木结构建筑的基本构件说起。它们一般包括柱、梁、枋、檩条、斗拱、椽子、望板等。这些构件组合起来才能建成房屋。中国传统木结构建筑采用的是榫卯连接，就是将两个构件的凹凸部位嵌合在一起，这样的构件可以预制加工，现场装配。

小　龙：好像榫是突出来的，卯是凹进去的，对吗？

林教授：是的。像太和殿这样大型的木结构宫殿，需要成千上万的木构件。这些木构件，除了椽、望板这类屋面板材，其余的几乎全靠榫卯结构连接在一起。

小　龙：榫卯结构有很多种吧？

林教授：没错。榫卯结构种类繁多，各有各的功能，可以固定不同类型的构件，垂直的、水平的、倾斜的都可以，还可以用于板缝拼接。

大　卫：听起来，榫卯结构的功能很多啊。

林教授：是的。榫卯结构还有一个不同寻常的地方，就是有很强的柔性连接功能，因此建筑物的抗震能力较强。中国自古就有"墙倒房不塌"的说法。

小　龙：确实是。书上说，每个榫卯结点就像一个弹簧。在强震中，木结构建筑即使发生一定幅度的摇晃，有

斗拱
Dougong

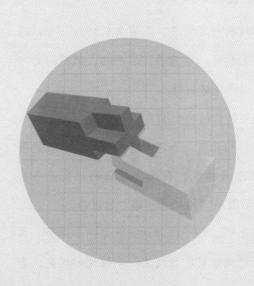

榫卯
Mortise-and-Tenon Joints

一定的变形，也不会轻易倒塌。

大　卫：还有抗震功能？也太先进了吧？这么好的榫卯结构只是用来建房屋吗？

林教授：当然不是。榫卯结构用处很多，可用在建筑中，也可用在车辆、船只、桥梁、农具、家具，甚至乐器上。

小　龙：我知道古典家具都是榫卯结构，不用钉子的。

大　卫：苏州园林里就有很多漂亮的中式家具，我还不知道都不用钉子呢。

转角斗拱　Corner *Dougong*

林教授：榫卯结构是中国古代物件连接的基本方法，凹凸自然对接，体现了阴阳互补的中国传统智慧。了解了榫卯结构，我们来好好看看屋檐下的这些斗拱吧。

大　卫：这些斗拱一排排好密集，看上去很复杂。

林教授：是的。你们猜猜看，太和殿上下两层檐一共有多少组斗拱？

大　卫：大概有200多吧？

林教授：不止200，一共有368组，加上殿内的136组……

大　卫：一共504组。

林教授：算得挺快啊。太和殿斗拱种类齐全，构造复杂，保存完好，是了解斗拱结构最好的地方之一。

小　龙：林教授，殿内外的斗拱一样吗？

林教授：不一样。在建筑物外檐部位的叫外檐斗拱，殿内的斗拱叫内檐斗拱。斗拱的种类很多，比如，外檐中结构最复杂的是转角斗拱，也是最漂亮的斗拱结构。

小　龙：这么多斗拱，结构上都有什么特点呢？

林教授：根据建筑规模不同，斗拱可分为三踩、五踩、七踩等不同的结构形式。建筑级别越高，斗拱踩数越大。

大　卫：这个太和殿是几踩斗拱？

林教授：九踩，是最高级别了。

大　卫：这些斗拱很漂亮，都有什么功能呢？

林教授：外檐斗拱的功能是承受上部支出的屋檐，将其重量或直接集中到柱上，或间接地先传到额枋上再转到柱上。内檐斗拱在室内承托天花枋，构成室内天花。

小　龙：那它们的功能就是传递重量荷载吗？

林教授：那只是斗拱四大功能之一。斗拱第二个功能是承托出檐，也就是承托屋檐伸出梁架之外的部分。第三个功能是标记建筑物的等级。第四个功能是用作度量单位。

小　龙：林教授，这个怎么用作度量单位啊？

林教授：就是说，斗拱中坐斗的开口尺寸是基本单位，其余构件的尺寸都是它的倍数。按照斗口尺寸，可以算出带斗拱的建筑各部位、各构件的详细尺寸。

小　龙：您能举个例子吗？

林教授：比如一栋建筑的斗口尺寸为10厘米，檐柱直径应该是6斗口，就是说檐柱的直径是60厘米。

大　卫：这下明白了。没想到斗拱还有尺子的功能！

林教授：是的，不过在早期建筑中，斗拱的主要功能还是承重。后来，它的结构性功能慢慢弱化，装饰作用增强。人们甚至会把斗拱的部件设计成动植物或器物的形状，和彩绘结合起来非常漂亮。我们进殿里看看，里面有更多的斗拱构件，还有和玺彩画和漂亮的藻井。

大　卫：好。

Mortise-and-Tenon Joints and *Dougong*

> Xiaolong, David, and Prof. Lin were standing at the entrance of the Hall of Supreme Harmony.

Prof. Lin: Xiaolong, David, let's go and take a closer look at the timber frame of this hall. Look! Those components above the architrave are ingenious.

Xiaolong: Are you talking about those interlocking brackets?

Prof. Lin: Yes. The interlocking bracket is called *dougong* in Chinese. They're exclusive to Chinese buildings. Back in the Spring and Autumn Period, they had already been used to enhance the stability and beauty of buildings. Here, in the Hall of Supreme Harmony, they are particularly elaborate and representative. To understand this component, we first need to know its structure.

Xiaolong: The mortise-and-tenon joint, right? A book I once read says that it's a joint without nails, but I didn't quite get it. How can we put all these components together without nails?

Prof. Lin: Let's start with the basic components of timber-frame buildings. Generally, they are columns, beams, joists, purlins, interlocking brackets, rafters, and roof boardings. We have to assemble these components to build a house. In Chinese architecture, the assembly method is to make the mortise-and-tenon joints. It means inserting tenons into mortises. Therefore, all components are pre-made with tenons and mortises so as to ensure quick assembly on site.

Xiaolong: Are tenons convex and mortises concave?

Prof. Lin: Yes. Large timber-frame buildings, like this hall, are composed of thousands of timber components. And all the components, except for rafters and roof boardings, are assembled mainly by the mortise-and-tenon joints.

Xiaolong: It seems that there're many kinds of mortise-and-tenon joints.

Prof. Lin: Yes, there're many categories with different functions. They can be used to fix different components, vertical, horizontal, or diagonal. Besides, they can be used to seal cracks between planks as well.

David: Wow. They seem to be multifunctional.

Prof. Lin: Yes. One more thing. The mortise-and-tenon joints are strikingly flexible. An architecture with these joints is highly anti-seismic. This is why we have an old saying, "A timber-frame building won't collapse even if its walls fall down."

Xiaolong: Indeed. I read that each mortise-and-tenon joint is like a spring. This is why in a strong earthquake, the timber-frame buildings would shake and suffer some deformation, but less likely collapse.

David: So these buildings are anti-seismic. Fabulous! Is this joint only used in constructing buildings?

Prof. Lin: Of course not. Besides architecture, it's widely used in many other things, such as vehicles, ships, bridges, farm implements, furniture, and even musical instruments.

Xiaolong: I know this kind of joint is used in classical Chinese furniture which is constructed without nails.

David: There're many pieces of classical Chinese furniture in Suzhou Gardens, but I didn't know there're no nails in them.

Prof. Lin: The mortise-and-tenon joint is a fundamental assembly method in ancient China. The natural clamp of convex and concave reflects the traditional Chinese philosophy, namely, the complementarity of *yin* and *yang*. OK, now let's take a closer look at these interlocking brackets, or *dougong* under the eaves.

David: They are densely placed under the eaves and look quite complicated.

Prof. Lin: Yes. Guess how many sets of *dougong* are there in the upper and lower eaves?

David: More than 200?

Prof. Lin: That's a conservative estimate. There are 368 sets. Besides, there are 136 sets inside the hall...

David: 504 sets in total.

Prof. Lin: Good for you, David! The Hall of Supreme Harmony has a full range of *dougong*, all ingeniously made and well-preserved. It's one of the best places to learn about *dougong*.

Xiaolong: Prof. Lin, are the sets of exterior *dougong* the same as the interior ones?

Prof. Lin: No, they aren't. Under the exterior eaves, they are called the exterior eave brackets, and these inside the hall are the interior eave brackets. There are many categories of *dougong*. For example, the corner *dougong* under the exterior eaves are the most complicated and most beautiful.

Xiaolong: What's the structural difference between these varieties of *dougong*?

Prof. Lin: According to the scales of different buildings, there are three-step *dougong*, five-step *dougong*, seven-step *dougong*, and so on. The higher rank the building holds, the more steps the bracket cluster possesses.

David: Well, how many steps do the interlocking brackets in the

Hall of Supreme Harmony have?

Prof. Lin: Nine, the highest level.

David: These interlocking brackets are very beautiful. What are their functions?

Prof. Lin: The exterior eave brackets are to support the eaves on them, transferring the weight either directly to the columns, or first to the architraves and then to the columns. And the interior eave brackets are used to support the ceiling joists inside the building and constitute the interior caisson ceiling.

Xiaolong: Their main function is weight transferring, right?

Prof. Lin: That's only one of their major four functions. Their second function is to support the exterior eaves, the part of the eaves that extends beyond the beam frame. The third one is to mark the rank of the building. And the last one is to serve as a measurement unit.

Xiaolong: Prof. Lin, I didn't get it. How can the *dougong* serve as a measurement unit?

Prof. Lin: Actually, it's the block mortise width of *dougong* that serves as the basic unit, and the sizes of other components are multiples of it. With the block mortise width, we can work out the size of every part or every component of the building.

Xiaolong: Would you please give us an example?

Prof. Lin: Of course. Say, the width of the block mortise in a building is 10 cm, and the diameter of an eave column is six times of that, then the diametre is 60 centimetres.

David: Oh, I got it. That's indeed a measurement unit. Wonderful!

Prof. Lin: Yes. In the early days, the main function of *dougong* was to bear weight. Later, the structural function gradually gave way to the decorative function. People even designed them into the shape of plants, animals or other objects, and painted them with coloured pattern. Let's get into the hall and take a look. There are many more interlocking brackets inside. Besides, there are some dragon or phoenix patterns, and beautiful caisson ceilings.

David: Great.

藻井

> 小龙、大卫和林教授来到太和殿内,迎面看到云龙纹宝座。

大　　卫：这里真是金碧辉煌啊,到处都是龙。

林教授：猜猜看,这里有多少条龙?

大　　卫：也许1000条吧。

小　　龙：应该不止1000条,外面已经是千龙吐水了,这里恐怕得有3000多条吧?

林教授：你们都说少了,大殿内外各种形式的龙共有1万6000多条。

大　　卫：真的吗?这么多啊。

小　　龙：看来的确是真龙天子宝殿啊。

林教授：你们看这个楠木宝座,这是故宫现存做工最讲究、等级最高的宝座。

小　龙：咦，这个宝座有点儿特殊，没有椅子腿，用的是底座。这也是须弥座吗？

林教授：是的，就是须弥座。

大　卫：这里到处都是金色的龙，连天花板上都是啊。

林教授：大卫，这不叫天花板，叫天花，宝座上方的天花叫蟠龙衔珠藻井。

小　龙：藻井？是因为它像个井吗？

林教授：是啊。藻井是一种高级天花，中间凹陷下去，形似水井，外沿可以有多种形状，比如方形、多角形或圆形。藻井一般用于最高级别的建筑，比如宫殿、寺庙正中或者宝座上方。

大　卫：这个藻井里面是圆的，外面是方的，那应该算什么形状呢？

林教授：这个有些复杂。我们分三层看吧。最外层是方井，中间八角井，里面是圆井。太和殿的藻井既有装饰性又有象征性。

小　龙：是象征天圆地方吗？

林教授：不仅是天圆地方，还有天人合一的寓意。

大　卫：这个藻井虽然很复杂，但很对称。

林教授：是的，这种对称设计很符合中国传统美学理念，平

云龙纹宝座与藻井
The Gilded Throne with Cloud and Dragon Patterns and the Caisson Ceiling

衡和谐。故宫建筑装饰繁而不乱，变化中有秩序和节律，尽显对称之美。

小　龙：我看介绍说，早期藻井的纹样很繁琐，很像井里的水藻，所以叫藻井，同时也取以水镇火的意思。

林教授：确实如此，这里面的寓意很丰富。

大　卫：中国文化里的这些寓意太有意思了。

林教授：你们看，这个金碧辉煌的大殿，除了金色的装饰，就是画在梁、枋、斗拱和天花上这些漂亮的和玺彩画。

小　龙：和玺彩画？

林教授：是的，和玺是清式彩画中最高级别的彩画。

大　卫：所以用在最高级别的宫殿里。

林教授：是的，你们知道吗？这些龙的画法都是有规定的。刚才宫殿门外额枋正中画的是二龙戏珠，梁上是行龙，天花上是坐龙。

大　卫：林教授，我特别想知道，这么有立体感的画是怎么画上去的？

林教授：这种画法叫沥粉贴金。用一个尖端有孔的管子，装有胶和土粉混合成的膏状物，按彩画图案描出隆起的花纹，上面涂胶后再贴上金箔，这样图案就有了立体感。

小　　龙：和蛋糕裱花一样呀。

林教授：哈哈，你这个说法很形象。你们要知道，彩画是中国古代建筑中重要的装饰手法，很常见。这三大殿的彩画叫金龙和玺，故宫其他场所还有其他种类的和玺彩画。

小　　龙：明白了。宫殿级别和功能不同，所以采用不同的彩画。

林教授：是的。和玺彩画还有个特点，就是从枋心开始向两端对称构图。颜色也有固定要求，比如说，斗拱多用蓝绿两色，周角要用金色线。

小　　龙：嗯，精巧的木结构和这些彩画结合得真完美。

Caisson Ceiling

> Prof. Lin, Xiaolong and David got into the Hall of Supreme Harmony. A gilded throne with cloud and dragon patterns caught their eyes.

David: This hall is indeed resplendent and magnificent. Dragon patterns are everywhere.

Prof. Lin: Would you like to guess how many dragons there are?

David: 1,000?

Xiaolong: Well, I think there are more. Now that there are "thousands of dragons spouting water" on the platform outside, I guess there're more than 3,000 dragons in this room.

Prof. Lin: Far more than that. Inside and outside the hall, there are more than 16,000 dragons in total.

David: It's incredible! So many dragons!

Xiaolong: Now I see why it is the hall for emperors, who saw themselves as the incarnation of dragons.

Prof. Lin: Look! This elaborately finished throne, made of the wood nanmu, is the symbol of supremacy in the Forbidden City.

Xiaolong: Hey, this throne is exquisite. It is supported by a pedestal rather than legs. Is it a sumeru pedestal as well?

Prof. Lin: Bravo, Xiaolong! It is a sumeru pedestal.

David: Hmm. Golden dragon patterns are everywhere, even on the ceiling.

Prof. Lin: Sure thing, David. The part above the throne, with the sculpture of dragons holding pearls in their mouths, is called a "caisson ceiling" or *zaojing* (藻井) in Chinese.

Xiaolong: *Zaojing*? Is it because it looks like a well, or *jing* (井) in Chinese?

Prof. Lin: Yes. It's a high-end ceiling, with the centre concave, shaped like a well. The outer edge can have a variety of shapes, such as square, polygon or circle. It's generally placed in the centre or above the thrones in the top-end buildings like palaces or temples.

David: This caisson ceiling is round inside and square outside. Then how should I define its shape?

Prof. Lin: It's a little complicated. Let's analyse it from three levels. The outermost edge is a square; the middle, an octagon; and the innermost, a circle. The caisson ceiling here is both decorative and symbolic.

Xiaolong: Is it a symbol of the round sky and the square earth?

Prof. Lin: More than that. It also implies the harmony between man and nature.

David: The caisson ceiling, though complex, is symmetrical.

Prof. Lin: Yes, the symmetrical design is in line with traditional Chinese aesthetics, advocating balance and harmony. The decoration of buildings here is complex but well-organised. It displays order and regularity along with the variation, showing the beauty of symmetry.

Xiaolong: According to the travel guide, early caisson ceilings had elaborate patterns that look like algae in the well water, so they were called *zaojing* in Chinese. The term also metaphorically refers to the place where water can be fetched for extinguishing fire.

Prof. Lin: Exactly. The implied meanings are rich.

David: Metaphors in Chinese culture are interesting.

Prof. Lin: Look! In this glorious hall, there are not only golden decorations but also colourful dragon or phoenix patterns on beams, joists, interlocking brackets, and caisson ceilings.

Xiaolong: Colourful dragon or phoenix paintings?

Prof. Lin: Yes. Dragon or phoenix paintings are most exquisite colourful paintings in the Qing Dynasty.

David: That's why they are used in imperial palaces.

Prof. Lin: You're correct. These dragons are painted according to specific regulations. For example, "two dragons playing with a pearl" is painted on the centre of architraves outside this hall; "dragon in action" is painted on the beams; and "dragon in rest" is painted on the caisson ceiling.

David: Prof. Lin, I'm wondering how to create depth in these paintings?

Prof. Lin: This method is called "powdering and gilding". In order to add depth to the painting, they need a pipe with a hollow tip to make raised patterns after filling the pipe with paste made of glue and powder. At last, they'll apply glue on the raised parts and gild gold foil.

Xiaolong: It reminds me of decorating a cake.

Prof. Lin: That makes much sense. Colourful painting is a common and important decorative technique in ancient Chinese architecture. In the three major halls, they are called "golden dragon paintings". In other halls or places, there're many different kinds of colourful paintings.

Xiaolong: I see. Patterns vary according to the specifications and functions of the palaces.

Prof. Lin: Yes. There's one more feature of the dragon or phoenix paintings. Craftsmen will start painting from the central parts of the joists. Then they'll draw the patterns symmetrically to the ends. Besides, there are rigid regulations for the use of colours. For instance, blue and green are the primary colours in painting *dougong* while golden yellow is dominant in painting the outline.

David: The timber-frame structure and these colourful paintings match perfectly.

柱和梁

林教授：我们再来看看支撑着这个大殿屋顶的梁柱吧。

大　卫：中间的柱子还是金色的呢。

林教授：是的。太和殿正中央的六根柱子叫蟠龙金柱，是用纯度为99.99%的黄金制成的金箔包裹的。

小　龙：看，好像是两种不同的金色呢。这些龙的图案也是突出的，是和玺彩画吗？

林教授：是的。采用深浅两种颜色的金箔，是为了突出上面的蟠龙图案。

小　龙：蟠龙金柱是不是规格最高的柱子？

林教授：是的。六根蟠龙金柱只有正殿才能设置，是皇权的象征。你们知道金色柱子旁边的那些柱子叫什么吗？

大　卫：红色的圆柱呀。

林教授：不，也叫金柱。

小　龙：怎么能都叫金柱呢？只有中间的六根是金色的呀。

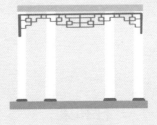

柱和梁　Column and Beam

林教授：木结构的柱子起名不是按颜色，而是按形状、位置或功能。它们叫金柱，是因为这些位置最重要。太和殿有六排三圈共72根木柱。中间这一圈叫内金柱。第二圈支撑重檐的也是金柱，叫重檐金柱。最外面的一圈是支持外檐的檐柱。

大　卫：啊，这六根金色的柱子原来还不是同一种柱子。

林教授：对，前四根柱子是内金柱，后面两根是重檐金柱。

小　龙：我觉得屋檐不要柱子支撑也可以吧？不用不是更好看吗？

林教授：还是需要的。传统木结构建筑中的屋檐都有出檐。越大型的建筑出檐越大，而这部分屋檐离屋顶的中心较远，单独的柱子支撑会分担一部分屋顶重量。

小　龙：还有其他功能吗？

林教授：有的，檐柱不仅用来支撑屋檐，同时也连接其他重要的木构件，比如斗拱、额枋。它们相互连接、共同支撑起整个屋顶。

大　卫：柱子必须是立在地面上的吗？

林教授：那还真不一定。关于柱子有两点需要注意。第一，有些柱子可以不落地；第二，虽然大部分柱子起承重作用，但也有例外。

小　龙：不落地也叫柱子？

林教授：有不少柱子是不落地的。还记得前面我们说过梁上的瓜柱吗？它们是重要的承重柱，但不落地。

大　卫：那不承重的柱子起什么作用呢？

林教授：主要是起装饰作用。你们看，两侧墙上的垂花门下面是不是有倒悬着的短柱？那个叫垂莲柱，就是起装饰作用。

大　卫：原来柱子还有这么多讲究，太长见识了！

小　龙：林教授，金柱上面就是梁吧？

林教授：是的，这里我们只能看到抬梁式结构中最下面的梁，那些短梁藏在天花上面。作为木建筑中最重要的部件之一，梁既承托建筑物的上部构件，比如枋、檩，还要负担屋面的全部重量。可以这么说，梁支撑着整个建筑物的上部。

小　龙：然后梁再把重量传递给柱子？

林教授：没错。如果建筑规模较小，梁可以直接放在柱头上。不过像太和殿这样的大型建筑，梁是放在斗拱上的。依据位置、形状和功能的不同，梁有各种不同的名称。你们看，我们现在面朝宝座，头顶上纵向的是天花梁。横向的是什么呢？

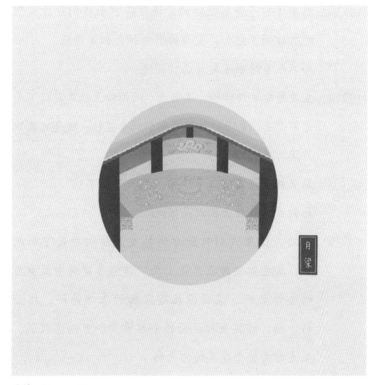

月梁　Crescent Beam

大　卫：那应该是枋吧？

林教授：是的，是金枋，是位于金柱与金柱之间的枋。

大　卫：我明白了，枋和梁的高度一致，但走向不一样。枋与梁垂直，与正脊平行。

小　龙：林教授，我发现南方的梁和这儿的不一样，不是直的，好像是弯弯的，我觉得那样的梁更漂亮。

林教授：那是月梁，确实很漂亮。北方的梁平直，南方的梁中段微微拱起，两端弯曲，很像弯弯的月亮。

大　卫：月梁，这名字好有诗意呀！像月亮的梁，我记住了。

林教授：月梁的侧面常雕刻精美的纹样或绘制漂亮的彩画。宋代以前，大型建筑基本上用的是月梁，后来北方渐渐不用了。

大　卫：这么漂亮的月梁，北方为什么不用了呢？

林教授：这是出于对气候的考虑。屋顶设计不仅要美观，更重要的是要实用。北方往往在梁架下多设一层天花板，使建筑物内部形成较小的闭合空间，有利于室内保温。南方天气不那么冷，保温不是主要问题，月梁露在外面，看起来并不突兀，反而显得建筑物有艺术气息。

小　龙：不同地区气候不同，木结构建筑也各有特色。

Column and Beam

Prof. Lin: Now look at the columns and beams that support the roof.

David: The columns in the centre are golden.

Prof. Lin: Yes. Those six golden columns with the pattern of coiling dragons are wrapped with pure gold foils.

Xiaolong: Hey, it seems that there are two different shades of golden colours. And these dragons are prominent. Are they also called dragon or phoenix paintings?

Prof. Lin: Yes, there are two shades of golden colours to highlight the coiling dragons.

Xiaolong: Do these columns rank the highest?

Prof. Lin: Yes. They were only used in the major halls because they symbolised the imperial power. Do you know the name of these columns next to the golden ones?

David: Red columns?

Prof. Lin: No, they're called "golden columns" as well.

Xiaolong: Wait, did you say "golden"? But only the middle six are golden.

Prof. Lin: The columns in the timber-frame buildings aren't named by their colours. Our ancestors named them by their shapes, locations or functions. They are called golden columns for their significant positions, precious as gold. There are 72 wooden columns here arranged in six rows forming three concentric circles. And those columns in the innermost circle are the interior golden columns. Those in the middle circle, supporting the lower eaves, are the lower-eave golden columns, and those in the outermost circle, supporting the exterior eave, are called the eave columns.

David: So the six golden ones are not the same kind?

Prof. Lin: No, the front four are the interior golden columns, while the other two are the lower-eave golden columns.

Xiaolong: Why do we need extra columns to support eaves?

Wouldn't eaves look better without those columns?

Prof. Lin: Actually, it is out of necessity. In traditional timber-frame buildings, there are overhanging eaves. The size of the overhanging eaves is in direct proportion to the building scale. In a massive building, the overhanging eaves will be far away from the centre of the building. They need extra columns to support them.

Xiaolong: Is that the only function of eave columns?

Prof. Lin: No. Besides supporting eaves, they are also connected with other important timber-frame components, like interlocking brackets and architraves. These interconnections help prop up the whole roof together.

David: Do all the columns stand on the ground?

Prof. Lin: Not really, David. About columns, there are two things worthy of mentioning. First, some of them don't stand on the ground; second, although most of them are load-bearing, there are exceptions.

Xiaolong: May they be called columns if they don't stand on the ground?

Prof. Lin: Yes. There are many columns that don't stand on the ground. Do you remember the dwarf columns we talked about before? They support weight, but they don't stand on the ground.

Xiaolong: What is the function of columns that don't bear load?

Prof. Lin: They are mainly used for decoration. Look at the festooned gates on the flank walls. Do you see the short columns that hung upside down? They are the lotus-festooned columns, used only for decoration.

David: I didn't expect so many regulations about columns. I do have learned a lot.

Xiaolong: Prof. Lin, above the golden columns are beams, right?

Prof. Lin: Right, but we can only see the beams at the bottom of the post-and-beam construction. Other short beams are hidden above the ceilings. As the most significant components in timber-frame architecture, beams not only support the components in the upper part of the building, like joists and purlins, but also bear the weight of the roof. We can say the whole upper part of the building is supported by the beams.

David: And then beams will transfer the load to columns?

Prof. Lin: Yes. If the building is small, beams can be placed directly on the columns. But in a large building, like this hall, beams are placed on the interlocking brackets. They are named according to their positions, shapes or functions. Look! Now, when we face the throne seat, those vertical components above our heads are the ceiling beams. Do you know what those horizontal ones are?

David: Joists?

Prof. Lin: Yes, they are golden joists, placed between the golden columns.

David: I see. The height of joists and beams is the same, but the direction is different. Joists are perpendicular to the beams and parallel to the main ridge.

Xiaolong: Prof. Lin, I find that the beams in southern China are not the same as these here. Those are not straight, but curved. I think they're more beautiful.

Prof. Lin: Those in the south are called crescent beams, very beautiful. In northern China, beams are flat and straight, while in southern China, beams are slightly arched in the middle and curved at the ends, like crescents.

David: Crescent beams? What a poetic name! It seems that the beam looks like the crescent. I got it.

Prof. Lin: The sides of crescent beams are often carved with exquisite patterns or painted with beautiful coloured patterns. Before the Song Dynasty, they were commonly used in large buildings, but later, they were barely used in northern China.

David: They're beautiful. I wonder why they were not used in northern China later?

Prof. Lin: Roof design has to adapt to different climates. In the design of roofs, practicability outweighs beauty. In the north, there is an extra ceiling under the beam frame, so that the interior of the building becomes an enclosed space. It's helpful for thermal insulation. In southern China, however, the weather is not as cold as that in the north, so thermal insulation is not the primary concern. Thus, the uncovered crescent beams come into being. Instead of being obtrusive, they actually add an aesthetic aura to the building.

Xiaolong: I see. Weather also accounts for the distinctive style of timber-frame buildings in different regions.

收分

> 出了太和殿,林教授提醒小龙和大卫注意观察大殿外檐的柱子。

林教授: 你们看一下,这些柱子上下一样粗吗?

大　卫: 一样啊。

小　龙: 我也没看出来不一样。还能不一样粗吗?

林教授: 肉眼看不出来,但它们确实上下不一样粗。柱子下粗上细,这种设计叫收分。

小　龙: 收分有什么具体规定呢?

林教授: 一般收分的原则是减去柱高的1%。举个例子,假设一栋房子的柱子高5米,那么收分应该是5厘米。大型建筑的收分通常是柱高的0.7%。

小　龙: 怪不得看不出来呢。我还有一个问题,资料里说檐柱应该是往里斜一些,我也没觉得斜呀。

大　卫: 对啊,这不是很直吗,不斜呀。

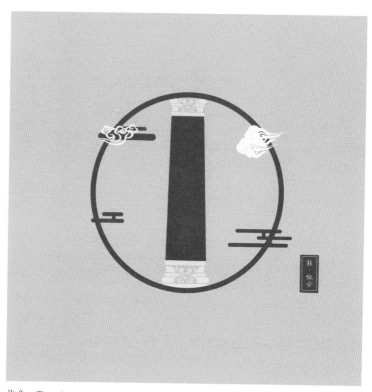

收分　Tapering

林教授：没错。这个斜度肉眼是看不出来的。这是另一项技法，叫侧脚，就是将建筑物外圈柱子的底部向外侧移一点儿，使上端略向内侧倾斜。侧脚的尺寸比例与收分相同。

大　卫：明白了。就是说柱高要是10米，收分7厘米，侧脚也是7厘米。

小　龙：为什么要收分和侧脚呢？

林教授：这个问题问得好。你们说建筑最重要的是什么？

大　卫：结实稳定呗。

林教授：非常正确。侧脚可以形成在顶点虚拟相交的三角形结构，收分有利于上部重量顺利传递到地面。这两种技术都是为加强建筑物的稳定性。

小　龙：真没想到，普通的柱子里还蕴含着这么多智慧！

Tapering

> Walking out of the Hall of Supreme Harmony, Prof. Lin asked Xiaolong and David to observe the columns under the exterior eaves.

Prof. Lin: Look! Are the columns under the exterior eaves of the hall the same in size from top to bottom?

David: Yes, I think so.

Xiaolong: Well, I think so as well.

Prof. Lin: Surely you can't tell through naked eyes. In fact, their thickness is not the same from top to bottom. Here the tapering technique or the bottom-to-top contraction is used.

Xiaolong: Is there a rule for it?

Prof. Lin: Of course. For most columns, the ratio of contraction is 1% of the column height. For example, for a building with five-metre-high columns, the contraction should be five centimetres. For some large buildings, the ratio is 0.7%.

Xiaolong: No wonder I can't tell. And I have one more question. Books say that the eave columns usually incline inwards slightly, but I can't tell that either.

David: Incline? It seems that these columns are perfectly vertical.

Prof. Lin: Yes. You can't tell that through naked eyes either. It's another traditional technique. The bottoms of the outer columns incline outwards a little, so that the tops of these columns are slightly inclined inwards. The ratio of inclination is the same as that of the column tapering.

David: I got it. That is to say, if the height of a column is ten metres and the contraction is seven centimetres, then the bottom of the column should also be budged outwards seven centimetres.

Xiaolong: But why do we need these two techniques?

Prof. Lin: Good question. What do you think is the most important thing about architecture?

Xiaolong: I think it's sturdiness and stability.

Prof. Lin: Right. The inclination can make the columns a triangular structure that virtually intersects above the building, while the column tapering is conducive to load transfer. These techniques are used to enhance the stability of buildings.

Xiaolong: Wow, fancy so much wisdom behind these common columns.

攒尖顶

出了太和殿后门,小龙、大卫和林教授来到中和殿前。

林教授:中和殿是供皇帝休息的地方。在太和殿举行大典之前和大典期间,皇帝都会在中和殿休息。你们看,中和殿屋顶有没有特别之处?

大　卫:看上去像亭子。不过比苏州的亭子大多了。

小　龙:没有正脊,只有一个尖顶!

林教授:没错。中和殿用的是攒尖顶。你们看,建筑物的屋面在顶部交汇成一点。攒尖顶是中式亭子最常用的顶,所以大卫觉得中和殿看起来像座亭子。

大　卫:原来是这样。

林教授:天坛的祈年殿也是典型的攒尖顶。

小　龙:就是天坛公园里的那个蓝色圆顶的大殿吗?我们昨

攒尖顶 Pyramidal Roof

中和殿 The Hall of Central Harmony

天刚去过天坛。可是那个屋顶和这个一点也不一样啊。这个有四个角,那个屋顶像个大伞。

林教授: 看上去很不一样,但它们都是攒尖顶。中和殿是单檐四角攒尖顶,有四条垂脊,正中是金顶。祈年殿是三重檐圆形攒尖顶,正中也是金顶。圆形攒尖顶没有垂脊,所以看上去很不一样。

大　卫: 攒尖顶就这两种吗?

林教授: 可不止两种,类型很多。中式园林中的亭子最常用攒尖顶,有三角、四角、五角、六角、八角等。此外,不同形式的攒尖顶组合起来,还能形成更加多样化的屋顶类型。如果再加上重檐甚至三重檐的变化,那么屋顶的类型就更丰富了。

小　龙: 攒尖顶居然能变换出这么多花样。

林教授: 是的。攒尖顶建筑既可以像中和殿、祈年殿一样雄伟壮观,也可以像苏州园林里的亭子一样精致优雅。

Pyramidal Roof

> Out of the back door of the Hall of Supreme Harmony, Xiaolong, David, and Prof. Lin went to the front of the Hall of Central Harmony.

Prof. Lin: Emperors used to rest in this hall before and during the ceremonies held in the Hall of Supreme Harmony. Look at its roof. Is there anything special?

David: Well, the roof looks like that of a pavilion, but it's much bigger than those in Suzhou Gardens.

Xiaolong: The roof has no principal ridge but one spire!

Prof. Lin: You're right. It is a pyramidal roof. The slopes of the roof meet at the top to form a spire. This kind of roof is very common for Chinese pavilions. That's why David thought this hall looked like a pavilion.

David: I see.

Prof. Lin: The roof of the Hall of Prayer for Good Harvest in the Temple of Heaven is also a typical example.

Xiaolong: Is that the big hall with a blue dome in the Temple of Heaven Park? We just visited it yesterday. Its roof is not the same as this one. It looks like a big umbrella, but this one has four corners.

Prof. Lin: Actually, both of them are pyramidal roofs. The roof of the Hall of Central Harmony has four diagonal ridges with a single eave intersecting at the golden spire. The roof of the Hall of Prayer for Good Harvest, though with a golden spire as well, is a round pyramidal roof with triple eaves. It has no diagonal ridge, so it looks very different from that of the Hall of Central Harmony.

David: So there are only two kinds of the pyramidal roof, right?

Prof. Lin: No, there are many other types, such as the trigonal roof, the tetragonal roof, the pentagonal roof, the hexagonal roof, and the octagonal roof. Besides, the different combinations of these roof types can form more other new types. And if they are built with double eaves or triple eaves, they can be even more diversified.

Xiaolong: Wow! It's beyond my expectation.

Prof. Lin: With different kinds of the pyramidal roof, buildings can be as majestic as the Hall of Central Harmony and the Hall of Prayer for Good Harvest, or as delicate and elegant as the pavilions in Suzhou Gardens.

歇山顶　悬山顶　硬山顶

> 出了中和殿，小龙、大卫和林教授来到三大殿的最后一个殿——保和殿。

大　卫：林教授，这个屋顶好像又是一种新样式？

林教授：是的，保和殿的屋顶是重檐歇山顶。故宫里还有两个重要的门也是用的重檐歇山顶。

大　卫：哪两个？

林教授：天安门和太和门。

小　龙：这个歇山顶的屋脊很多呀。一、二、三……八、九，一共有九条，对不对？

林教授：没错！上半部分有一条正脊和四条垂脊，下半部分四角上各有一条戗脊，一共是九条脊，所以歇山顶又称九脊顶。叫它歇山顶是因为正脊两端到屋檐中间断开一次，好像歇了一歇。

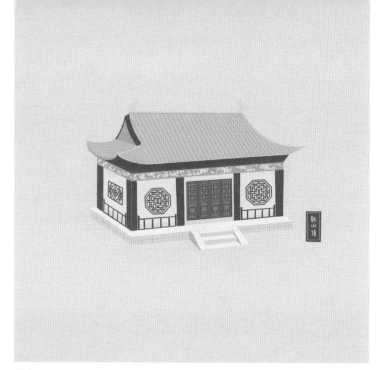

歇山顶 Gable and Hip Roof

保和殿歇山顶 The Gable and Hip Roof of Hall of Preserving Harmony

歇山顶 悬山顶 硬山顶
Gable and Hip Roof, Overhanging Gable Roof and Flushed Gable Roof

大　卫：这个名字很形象呀。

林教授：是的。歇山顶上面的那部分有前后两面坡，下面的部分有前后左右四面坡，分属两种不同的屋顶形式。

小　龙：我看出来了，歇山顶下面的部分是庑殿顶，上面的是什么形式呢？

林教授：这叫悬山顶，它在等级上低于庑殿顶和歇山顶，是民间建筑中最常见的屋顶形式。

大　卫：故宫里有专门的悬山顶宫殿吗？

林教授：正殿一般不用悬山顶。有些宫殿的配殿会用悬山顶。中国建筑里常见的屋顶式样在故宫里都能看到。我们先进保和殿里面看看。后面能看到悬山顶的。我考考你们，看看是否记住了我前面讲过的东西。你们看这个大殿是几间呀？

小　龙：我得数一下，正面看十根柱子，侧面看六根柱子，那就是面阔九间，进深五间。

林教授：是的。再看上面的天花，这是什么彩画？

大　卫：和太和殿一样，是和玺彩画。

林教授：不错，不错。看来你们都记住了。

大　卫：这个殿看上去很宽敞呀。好像柱子少了一些。

歇山顶侧面　Profile of the Gable and Hip Roof

林教授：你的感觉不错。保和殿的柱子比常规的要少，所以感觉空间很宽敞。

小　龙：柱子少了，房屋的安全性会不会差一些？

林教授：不用担心，在木结构建筑营造中，减和增都是有规矩的。

> 小龙、大卫和林教授一行走出保和殿。夕阳照耀下，歇山顶侧面的金色图案非常漂亮。

大　卫：你们看，多好看呀。

李教授：这个侧面叫山墙，上面的画叫山花。

小　龙：名字起得真好。这个歇山顶大方漂亮。

林教授：还有更好看的歇山顶呢。等参观完，出故宫后，你们一定得在护城河外面看看角楼，那是我最喜欢的故宫建筑。角楼倒映在晚霞中的护城河里，特别美。

小　龙：角楼也是歇山顶？

大　卫：你是说那个传说里鲁班启发师傅们建造的角楼吗？

林教授：是的，就是那四个角楼，用的也是歇山顶，不过那个更复杂一些，上下两层有5个屋顶，结构精巧，特别富有艺术性。角楼是紫禁城的标志性建筑。

悬山顶　Overhanging Gable Roof

> 出了保和殿，大卫、小龙和林教授三人离开中轴线，来到西线上的慈宁宫。大卫一进门就注意到侧殿的悬山顶屋顶。

大　　卫：你们看，这不是悬山顶吗？

林教授：大卫眼睛很尖呀。这个侧殿的屋顶就是悬山顶。

小　　龙：好像慈宁宫的正殿规格很高呢，也是重檐歇山顶呀。

林教授：我们前面看的三个大殿是皇帝办公用的。现在我们来到的是后宫，皇帝和家人生活的地方。这里用重檐歇山顶，因为这里是太后、太皇太后住的地方。你们谁来说一下这个侧殿屋顶的特点？

大　　卫：屋脊是前后两面坡、五条脊，是吗？

林教授：是的，一条正脊和四条垂脊。还有一个特点，你们发现了吗？

小　　龙：屋脊的侧面出来了一点儿，没有和墙齐平，这能算特点吗？

林教授：就是这个特点。悬山顶的侧面墙往里缩进了一点儿，梁不露在外面。

小　　龙：好像悬山顶和另外一种屋顶很像，也叫什么山顶的。

林教授：你是说硬山顶吧。硬山顶和悬山顶都是前后两面坡，相似度很高，区别在于硬山顶的梁架全部封砌在山墙内，而悬山顶的梁架是悬出山墙之外的。等会儿去看西六宫，侧殿都是硬山顶的。

小　龙：哦，原来这里"山"是"山墙"的意思。刚才我还纳闷屋顶和山有什么关系？

林教授：有关系的是墙，不是山。可不要小瞧悬出的这一部分。屋顶延伸出山墙之外，能够起到防止雨水侵蚀墙体的作用，所以在多雨的南方，民居中常用悬山顶。而在干燥的北方，硬山顶房屋更常见。有空你们去看看北京四合院，里面大多是硬山顶的房屋。

大　卫：硬山顶的好处是什么？

林教授：硬山顶的山墙与屋顶齐平或略高，可以防火、保温。

小　龙：看来硬山顶的确更适合北方的气候特点。

林教授：悬山顶比硬山顶看上去灵巧，但缺点是上面一部分没有被包裹在墙内，有些木构件暴露在外，容易朽烂。

小　龙：看来不同形式的屋顶各有优缺点。

林教授：没错。庑殿顶、歇山顶、攒尖顶、悬山顶、硬山顶，这是传统木结构建筑中最常见的五种屋顶。它们和不同的屋檐搭配，形成形态各异、等级不同的屋顶结构。

大　卫：这故宫建筑又有技术，又有文化，太有趣了。

林教授：我们今天只是走马观花看个大概。感兴趣的话，以后可以再找时间仔细了解。明天我们去看另一个中国特色的建筑——牌楼。

小　龙：我猜，我们要去看最有名的"正阳桥"牌楼，对吗？

林教授：对的，正是。

Gable and Hip Roof, Overhanging Gable Roof and Flushed Gable Roof

Out of the Hall of Central Harmony, Xiaolong, David, and Prof. Lin went to the Hall of Preserving Harmony.

David: Prof. Lin, this roof seems to be a new category.

Prof. Lin: Yes. The Hall of Preserving Harmony adopts a gable and hip roof with double eaves. There're two important gates using this kind of roof in the Forbidden City.

David: What're they?

Prof. Lin: Tian'anmen and the Gate of Supreme Harmony.

Xiaolong: Look! There're many ridges on this roof. Let me count. One, two, three... eight, nine. It's nine.

Prof. Lin: Correct! In the upper part, there's a main ridge and four diagonal ridges. There're four propping ridges in the lower part. There're nine ridges in total, so this type of roof is also known as "a nine-ridge roof". From each end of the four diagonal ridges to the eave, there are four propping ridges. It seems that four diagonal ridges "rest" in the middle.

David: What a vivid metaphor!

Prof. Lin: Yes. The upper part has only two slopes, the front and the back, while the lower one has four slopes, the front, the back, the left, and the right. These two parts belong to two different types of roof.

Xiaolong: I know. I can see that the lower part is a hip roof, but I have no idea about the upper part.

Prof. Lin: It's an overhanging gable roof, which is lower in rank than the hip roof or the gable and hip roof. It's commonly used in dwellings for ordinary people.

David: Is there a palace with an overhanging gable roof in the Forbidden City?

Prof. Lin: Yes. Generally, in major halls, we won't see this kind of roof, but it is in some side halls. In the Forbidden City, we can see all kinds of roofs. Let's go into the

Gable and Hip Roof, Overhanging Gable Roof and Flushed Gable Roof

              Hall of Preserving Harmony. Here, I want to give you a quiz to see if you have mastered what I told you. How many bays are there in this hall?

Xiaolong: Let me count. There are ten columns in the front and six in the side. Thus, the width of the building is nine bays, and the depth five bays.

Prof. Lin: Correct. And look at the caisson ceiling above. What kind of coloured patterns are they?

David: They're the same as the ones in the Hall of Supreme Harmony. They are dragon or phoenix patterns.

Prof. Lin: Good!

David: How spacious this hall is! And the columns here are fewer, right?

Prof. Lin: Sure thing. The columns in this hall are fewer, so it's spacious.

Xiaolong: But is it safe with fewer columns?

Prof. Lin: Don't worry. In timber-frame architecture, there're rigid regulations about the number of columns.

> Xiaolong, David, and Prof. Lin walked out of the Hall of Preserving Harmony, and saw that the golden patterns on the side of the gable and hip roof were very beautiful in the setting sun.

David: Look at these patterns. How beautiful they are!

Prof. Lin: The side is a gable wall. And the patterns on it are called pediments or *shanhua* (山花) in Chinese.

Xiaolong: What a good name! This gable and hip roof is elegant and beautiful.

Prof. Lin: There are more beautiful ones in the palace. When we finish this tour, you may appreciate the turrets from the moat bank outside the Forbidden City. They're my favourite and very beautiful when reflected in the moat under the evening glow.

Xiaolong: So these turrets also adopt gable and hip roofs, don't they?

David: And are they the turrets mentioned in the legend where Lu Ban presented himself to inspire the craftsmen?

Prof. Lin: Yes. The turrets are more complicated by adopting the gable and hip roofs. On the top of each turret,

歇山顶 悬山顶 硬山顶
Gable and Hip Roof, Overhanging Gable Roof and Flushed Gable Roof

there are five intricate and artistic roofs. Now they're the landmarks of the Forbidden City.

> David, Xiaolong, and Prof. Lin left the Hall of Preserving Harmony, and went to the Palace of Compassion and Tranquility in the west. David noticed the overhanging gable roof of the side hall as soon as he entered the yard.

David: Look! An overhanging gable roof, right?

Prof. Lin: You're right, David. The roof of this side hall is an overhanging gable roof.

Xiaolong: Well, the roof of the main hall of the Palace of Compassion and Tranquility seem to be a gable and hip roof with double eaves as well. It must be of a high rank.

Prof. Lin: The previous three halls we visited were emperors' offices. Now we are in the empress' palace, also the residence of the emperor and his family. Because this palace was for the residence of the emperor's mother and grandmother, the roof is also a gable and hip roof with double eaves. Can you name the features of this side hall's roof?

David: Well, it's a roof with two slopes, the front and the back. Besides, it has five ridges, right?

Prof. Lin: Yes, you're right. One main ridge and four diagonal ridges. Is that all?

Xiaolong: The sides of the main ridge extend a little bit with some space between the wall and the eaves. Is this another feature?

Prof. Lin: Yes, that's it. The side walls under the overhanging gable roof retract a little bit, and the beams are not exposed to the outside.

Xiaolong: I remember that the overhanging gable roof is very similar to another kind of gable roof, but I can't recall it now.

Prof. Lin: That must be the flushed gable roof. These two kinds of roofs are very similar since both of them have the front and back slopes. The only difference is that the truss of the flushed gable roof is all built within the gable walls, while that of the overhanging gable roof extends over the gable walls. Later, we'll visit the six west palaces. All the side halls there have flushed gable roofs.

Xiaolong: So, gable refers to the gable walls. Just now, I was wondering why a roof is related to a gable.

Prof. Lin: Now, you got it. It's the wall that is closely related to the roof. Don't underestimate the overhanging part that extends beyond the gable wall. It can prevent rainwater from eroding the wall. Therefore, in rainy southern China, the overhanging gable roofs are more common in dwellings, while in arid northern China, the flushed gable roofs are more common. If time permits, you can visit some quadrangle dwellings in Beijing. Most of them have the flushed gable roofs.

David: What are the advantages of the flushed gable roofs?

Prof. Lin: Their gable walls are even with or slightly higher than the roof, which is helpful for fire prevention and heat preservation.

Xiaolong: It seems that the flushed gable roofs are more suitable to the climate in northern China.

Prof. Lin: It's true. The overhanging gable roofs are more delicate than the flushed gable roofs, but their deficiency is that some timber components are exposed, so, they're prone to decay.

Xiaolong: It seems that different kinds of roof have their own strengths and weaknesses.

Prof. Lin: Exactly. The hip roof, the gable and hip roof, the pyramidal roof, the overhanging gable roof, and the flushed gable roof are the most common roofs for the traditional timber-frame architecture in China. They form varied structures and mark different ranks with different kinds of eaves.

David: We can enjoy both technology and culture in the buildings of the Forbidden City. So interesting.

Prof. Lin: Today, we just took a brief tour here. If you're interested, you can get around to learn more next time. Tomorrow, we will see another kind of building with a Chinese style, the archway.

Xiaolong: Are we going to visit the famous Zhengyang Bridge Archway?

Prof. Lin: Yes, that's it.

木牌楼

> 第二天,小龙、大卫和林教授来到前门大街上的正阳桥牌楼前。

林教授:瞧,这就是正阳桥牌楼,也叫"五牌楼"。这是2008年照原样在原地复建的。

大　卫:太漂亮了。

林教授:牌楼是老北京街上的一大特色。常规的有木枋、石枋和琉璃枋牌楼,近代还有混凝土牌楼。这个五牌楼是老北京最有名的牌楼。

大　卫:听说北京是中国牌楼最多的城市?

林教授:是的。最能体现老北京风格的三样东西是胡同、四合院和牌楼。原来北京有300多座牌楼,现在只剩下65座了。

大　卫:在中国,牌楼的历史很长吗?

木牌楼　Timber Archway

正阳桥牌楼　The Zhengyang Bridge Archway

林教授: 这种建筑最早出现在周朝，一开始用于祭天。后来牌楼用于纪念孔子、表彰节妇孝女，再后来扩大到纪念名人大事、标志街巷区域的界限，还用于增加主体建筑的气势。

小　龙: 这么多作用啊。这复建后的正阳桥牌楼应该是木结构吧？

林教授: 是的。知道它为什么叫五牌楼吗？

大　卫: 是因为这五个额枋像牌子吗？

林教授: 这样理解也行。更确切地说，牌楼可以按照间数和楼数进行细分。和房屋一样，牌楼的间是柱子之间的空间，两柱为一间，一个屋顶叫一楼。正阳桥牌楼是六柱五间五楼，所以就叫五牌楼。

大　卫: 上面一层额枋上还雕着二龙戏珠呢。

小　龙: 旁边也有龙，好像是行龙呢。上面是斗拱。

大　卫: 上面是庑殿顶吧。

林教授: 真不错，你俩把昨天在故宫学到的都用上了。这个牌楼还有个特点，你们看看柱子和屋顶？

小　龙: 柱子高出屋顶了？

林教授：对。传统木牌楼的样式主要分柱出头式和柱不出头式两种。柱出头式木牌楼的立柱高出檐楼，这个就是柱出头式；立柱不超出檐楼的那种更常见。

大　卫：我喜欢这个牌楼的镂空雕花，很漂亮。

林教授：这可不仅是为了漂亮，它还能减轻不少重量呢。

小　龙：这些石头底座是为了更稳固吗？

林教授：是的，这些石头叫夹杆石，它们包裹着柱子，然后深埋地下，就是为了稳固牌楼。原始的木牌楼，夹杆石上应该有铁箍，有戗杆支持木柱。明天咱们去山西参观应县木塔，会看到更古老的传统木牌楼。

大　卫：太好了。

Timber Archway

> The next day, Xiaolong, David, and Prof. Lin went to the Zhengyang Bridge Archway on Qianmen Street.

Prof. Lin: Look, this is the Zhengyang Bridge Archway, also known as the Five-Bay Archway. This one is a replicate built according to the original at the same place in 2008.

David: It's so beautiful.

Prof. Lin: The archway is a distinctive feature of Beijing streets and roads in ancient times. Generally, there are three kinds of archways in terms of building materials, namely, timber archways, masonry archways, and glaze archways. In modern times, cement archways are added to the list. Among them, the Zhengyang

Bridge Archway is the most famous.

David: I heard that Beijing has the largest number of archways in China.

Prof. Lin: Yes. In the past, lanes, quadrangle dwellings, and archways are a common sight in Beijing. There were more than 300 archways in this city, but only 65 are left now.

David: Do archways have a long history in China?

Prof. Lin: Yes, of course. The archway was first used in the Zhou Dynasty to offer sacrifice to heaven. They were also used to commemorate Confucius or honour chaste women or filial daughters. Later, their functions expanded to commemorating famous people and major events, bordering blocks, and increasing magnificence of main buildings.

Xiaolong: Wow, fancy so many functions! This archway is made of timber, right?

Prof. Lin: Yes. Do you know why it's also called the Five-Bay Archway?

David: Isn't it because of these five architraves?

Prof. Lin: Well, not exactly. To be more precise, archways can be subdivided according to the number of bays and roofs. Like common buildings, the space between two columns in an archway is called a bay. There are six columns, five bays and five roofs of different heights in the Zhengyang Bridge Archway, so it's also called the Five-Bay Archway.

David: Look! The top architrave is carved with the pattern of "two dragons playing with a pearl".

Xiaolong: There are other dragon patterns as well. And above the architraves are interlocking brackets.

David: And on top of the brackets are hip roofs.

Prof. Lin: Pretty good. You two have shown what you have learned in the Forbidden City yesterday. But there's one more feature of this archway. Can you find it?

Xiaolong: The columns here are higher than the roofs.

Prof. Lin: Well said. Traditional timber archways are roughly divided into two types: archways with columns higher than the roof and archways with columns lower than the roof. This archway is the first type, although the second type is more common.

David: I like the carved patterns of this archway. It's very beautiful.

Prof. Lin: Well, it's not only for decorative purposes, but also for weight reduction.

Xiaolong: Are these stone bases used to stabilise the whole archway?

Prof. Lin: Yes, these stones are clamping stones, wrapping around the columns. They are buried deep to enhance the stability of the archway. In ancient timber archways, we can see iron hoops fixed on the clamping stones, and columns backed up by bracing poles. Tomorrow we'll go to Shanxi Province to visit Yingxian Wooden Pagoda. We will see more ancient timber archways there.

David: That would be great!

应县木塔

> 小龙、大卫和林教授来到山西应县木塔参观。他们看到大街上高耸的木牌楼、佛宫寺天王殿和巍峨的木塔。大卫和小龙围着木牌楼非常认真地观察。

林教授：你们仔细看看，它和北京的五牌楼有什么不一样的地方？

大　卫：林教授，这些斜着的木杆就是戗杆吧？哎，这些夹杆石上面还真有铁箍呢。

小　龙：这应该是悬山顶四柱三间三楼木牌楼吧。还有一点儿不一样的，这个是柱不出头式。

林教授：你们说得都很准确。

大　卫：我喜欢这些斗拱，太有气势了。

小　龙：你们看，那座木塔的塔顶好像是攒尖顶。有点儿远，看不出来是几角。

应县木塔　Yingxian Wooden Pagoda

三人来到木塔前,大卫跑上台基,围着木塔迅速跑了一圈。

大　卫:小龙,我数了一下,是八角形。应该是八角攒尖顶。这个塔可真大呀。

林教授：确实很大，光塔基直径本身就超过30米。大卫，小龙，你们了解这座塔吗？

小　龙：这个塔建于公元1056年，是辽代的建筑，是世界上现存最高、最大的木佛塔，全部是榫卯结构，正面应该是九开间。

林教授：说得没错。应县木塔是现存规模最大、最高，也是最古老的木结构楼阁式佛塔，俗称应县木塔，正式的名字叫佛宫寺释迦塔，因为这里面供的是佛祖释迦牟尼[①]。

大　卫：不是亲眼所见，真不能相信这么高的塔全部是用木头建成的。

小　龙：这塔真的好高。感觉有20层楼那么高。

林教授：差不多。准确高度是67.31米，相当于20多层楼的高度。

小　龙：好像一共有六层？

大　卫：不对吧？你看，第一层和第二层塔檐离得很近，应该是像故宫一样的重檐，上边四层应该是单檐。林教授，我说的对吗？

林教授：完全正确。准确地说，叫重檐五层塔。实际上每两层之间还有一个隐藏的楼层，所以一共有九层。

小　　龙：为什么要建隐藏的楼层呢？

林教授：隐藏的四个楼层能够很好地约束塔身，防止变形。

小　　龙：所以才千年不倒。

大　　卫：林教授，您看，虽然塔上的油漆都剥落了，但屋檐下的这些斗拱仍像花朵一样漂亮。

林教授：是的。木塔的斗拱造型被称为"百尺莲开"。等转到侧面，你们会看到那个"百尺莲开"的匾。

大　　卫：刚才我光顾着数有几个角了，没仔细看匾上写的什么。

林教授：斗拱是应县木塔的一大特色。这里被称为中国的斗拱博物馆。据专家考察，整个塔有斗拱54种、480朵。

大　　卫：这里的斗拱是几踩？

林教授：七踩，或者更确切地说是七铺作。

大　　卫：怎么又改说法了呢？

林教授：是这样的，木结构的很多名称在清朝前后有所变化。"踩"在清朝之前叫"铺作"。这是辽代的塔，所以叫铺作更准确。你们看，这些斗拱既提升了木塔的稳定性，又赋予它艺术性。

小　　龙：这么说,隐藏楼层加斗拱是这个木塔千年不倒的原因吧?

林教授：小龙,你忘了让木塔稳定的最重要元素了。

大　　卫：是柱子吧?

林教授：没错。设计师专门设计了内外两层环柱。每层的外围有一圈柱子,里面还有一圈柱子。两圈柱子把一层分成三部分,里圈中心部分是内槽,里外两圈柱之间是外槽,最外面是挑出的回廊。

大　　卫：不对呀,塔里有两圈柱子,外面还有柱子呢,应该是三圈柱子。

林教授：是这样的,只有底层外加了24根副阶柱。塔的其他各层都是内圈8柱、外圈24柱。

大　　卫：明白了。我算算啊,那一共有184根柱子。

林教授：算上短柱,有650多根呢。整个塔是用10万多个木构件建成的。

大　　卫：太厉害了。

小　　龙：这里面好狭窄呀,还很暗呢。现在已经禁止游客登塔,但我查到的信息上说这个塔曾经可容1000多人同时上塔,这怎么可能呢。

林教授：你的信息没错。这个塔可以容下1500人。底层很窄，中间有个高大的塑像。

小　　龙：到处都是斗拱。好像整个塔就是靠柱子和斗拱支撑起来的。

林教授：这是木结构的特点。你们知道吗？这个塔自重就2600多吨，按1500人同时在塔上计算，肯定超过3000吨，就是说每根柱子要承重100多吨呢。这些梁、柱、枋全靠木头之间的榫卯连接。

小　　龙：这些柱子也和故宫里的一样，向内倾斜吗？

林教授：是的，内外槽的柱子按照一定的角度倾斜，向内收拢，加强塔的稳固性。

大　　卫：里外上下都是斗拱，怪不得叫斗拱博物馆呢。

小　　龙：这里好像是抬梁式。你们看，有柱、有梁、有斗拱。还能看到塔檐上的檩。

林教授：不对。木塔斗拱上面有斜撑梁、木枋和短柱，它们组成不同方向的复梁式架构。抬梁式是用越来越短的梁一层层抬架上去，增加空间。而这种复梁式结构像一个圈，加强了木塔结构的整体性。这样既坚固，又美观。

小　　龙：将近1000年了，这个塔居然幸存下来，可真是不容易啊。

林教授：是啊，晴天的时候，10公里以外都能看见它。这个塔确实是久经沧桑，历尽风险。它经历过40多次地震，有的还是大震，遭受过无数次雷击。近代还经历了枪弹，遭受过炮击。现在塔身上还能找到弹孔和炮击的痕迹。

大　卫：这可真是个奇迹。一定得好好保护呢。

林教授：是的。现在这个塔开始倾斜了，但还没有找到好的解决方法。

大　卫：一定会有办法的。

小　龙：希望很快可以听到好消息。

林教授：会的。明天我们去悬空寺。看看那里的几根木头如何撑起一座寺庙。

注释：
① 释迦牟尼：古代印度佛教创始人。

Yingxian Wooden Pagoda

Xiaolong, David, and Prof. Lin were visiting Yingxian Wooden Pagoda in Shanxi Province. They saw a timber archway on the street. They also saw the Hall of Heavenly Kings of Fogong Temple in front of a lofty wooden pagoda. David and Xiaolong walked around the timber archway, observing it carefully.

Prof. Lin: Look carefully. What is the difference between this archway and the Five-Bay Archway in Beijing?

David: Prof. Lin, these slanted wooden poles are bracing poles, right? Wow, there indeed are iron hoops on these clamping stones.

Xiaolong: It must be a timber archway with four columns, three bays, and three roofs. It has an overhanging gable roof, and its columns are lower than the roofs.

Prof. Lin: You are right.

David: I like these interlocking brackets. They're imposing.

Xiaolong: The pagoda seems to have a pyramidal roof, but it's too far to tell how many angles there are.

> When they arrived at the pagoda, David sprinted up to the foundation and ran a full circle around the building.

David: Xiaolong, I did a quick count. There are eight angles. So it's an octagonal pyramidal roof. How grand it is!

Prof. Lin: Yes, it's grand. The base is more than 30 metres in diametre. David, Xiaolong, have you heard of this pagoda?

Xiaolong: Yes. It was built in 1056 in the Liao Dynasty. Now it's the highest and largest existing wooden pagoda in the world. All their components are connected through mortise-and-tenon joints. By the way, it's a nine-bay pagoda.

Prof. Lin: You're right. It's the largest, highest, and oldest existing wooden pagoda with multiple floors. It's commonly known as Yingxian Wooden Pagoda. However, its official name is Sakyamuni Pagoda of the Fogong Temple, for it's built to worship Sakyamuni[1].

David: I didn't believe that such a high pagoda could be purely made of wood until I see it now by myself.

Xiaolong: What a high pagoda! It seems to be as high as a 20-storey building.

Prof. Lin: Right. To be exact, it's 67.31 metres high, roughly equivalent to a building with more than 20 storeys.

Xiaolong: There seem to be six floors. Is that so?

David: I don't think so… Look! The first layer of eaves is very close to the second one. I think it is a roof with double eaves, just like the ones in the Forbidden City, and the upper four floors have single eaves. Prof. Lin, am I right?

Prof. Lin: Exactly, David! It's a five-storey pagoda with double eaves. But if you take a closer look, you'll find that there's a hidden floor between every two storeys. In fact, there're nine floors in total.

Xiaolong: Why did they build these hidden floors?

Prof. Lin: Because they can help shape the pagoda and prevent it from deformation.

Xiaolong: Maybe that's the reason why it can survive so many years.

David: Prof. Lin, look! Although the paint has peeled off, these interlocking brackets under the eaves are still very beautiful. They are like flowers.

Prof. Lin: Yes. People compare the bracket-structure of this pagoda to a blooming lotus. You can see a plaque of "Blooming Lotus" at its side.

David: Well, I only focused on the angles just now, not noticing the words on the plaque.

Prof. Lin: The interlocking brackets or *dougong* are major features of Yingxian Wooden Pagoda. This is why it's renowned as the museum of *dougong* in China. Experts say that there are 480 clusters of brackets in this pagoda, falling into 54 categories.

David: How many steps do the *dougong* here have?

Prof. Lin: Seven. Or to be precise, seven layers.

David: Why did you use the term "layer"?

Prof. Lin: Well, many terms for timber-frame architecture changed around the Qing Dynasty. For instance, the term for "step" was "layer" before the Qing Dynasty. This pagoda was constructed in the Liao Dynasty, so it is more accurate to use the term "layer". Look! These *dougong* not only enhance the stability of the pagoda, but also make it very artistic.

Xiaolong: So, the secret for its survival lies in its hidden floors and interlocking brackets.

Prof. Lin: Well, Xiaolong, you didn't mention the most important thing in keeping a wooden pagoda stable.

David: Oh, columns?

Prof. Lin: You're right. This pagoda has two rings of columns on each floor, an interior ring and an exterior ring. Each floor is divided into three parts in this way, the inner space inside the interior ring of columns, the space between the two rings of columns, and the overhanging corridor in the outermost part.

David: Wait a minute. You just mentioned that there are two rings of columns inside the building, but I see there are columns outside as well! There are three rings of columns, aren't there?

Prof. Lin: Well, here's the thing. Only the bottom floor was built with 24 attached corridor columns. Other floors in this pagoda have eight columns in the interior ring and 24 columns in the exterior ring respectively.

David: I see. Let me do the maths... That makes 184 columns in total.

Prof. Lin: Actually, there are more than 650 columns, including short columns. And there are more than 100,000 timber components in this pagoda.

David: Incredible!

Xiaolong: It's so narrow and dark here. It's a pity we can't go upstairs for its preservation. I read that this pagoda could accommodate more than 1,000 people at the same time. I just can't figure out how it's possible.

Prof. Lin: Sure. It can accommodate 1,500 people at the same time. The ground floor is very narrow because of this huge statue.

Xiaolong: Interlocking brackets are everywhere. It seems that the whole pagoda is supported by columns and interlocking brackets.

Prof. Lin: That's what makes timber-frame structures unique. Guess what! This pagoda weighs more than 2,600 tons. If it accommodates 1,500 people at the same time, the total weight would be more than 3,000 tons. In that case, each column would bear more than 100 tons. And all these beams, columns, and joists are connected by mortise-and-tenon joints.

Xiaolong: Do the columns incline inwards, just like those in the Forbidden City?

Prof. Lin: Yes. The two rings of columns are inclining inwards at a certain angle to strengthen the stability of the pagoda. Look at its interior structure.

David: Inside and outside, up and down, interlocking brackets are truly everywhere! No wonder it's called the museum of *dougong*.

Xiaolong: I think it's the post-and-beam construction. Look, there are columns, beams, and *dougong*. And we can also see purlins on its eaves.

Prof. Lin: It's not that simple. Look! Above the *dougong* are the diagonal supporting beams, timber joists, and short columns. They form a multi-beam construction in different directions. In the post-and-beam construction, beams of varing lengths are vertically arranged with the shortest one on the top. In this multi-beam construction, the whole structure is like a ring, strengthening the integrity of building. Hence, it's sturdy and beautiful.

Xiaolong: It is a wonder that this pagoda survived nearly 1,000 years.

Prof. Lin: Yes. On sunny days, you can see this pagoda from 10 kilometres away. It has indeed experienced all sorts of disasters, including more than 40 large earthquakes, and some of them were destructive. In recent times, it also suffered shooting and shelling. You can still find traces of bullet holes and shelling on it.

David: What a miracle! It deserves the best preservation.

Prof. Lin: Yes. Unfortunately, it's leaning now, and we have not yet found the best way out.

David: There must be some ways.

Xiaolong: I'm looking forward to good news.

Prof. Lin: I'm sure there will be. Tomorrow we'll go to the Hanging Temple to see how a few pieces of wood prop up a whole temple.

Note:

1. **Sakyamuni:** The founder of Buddhism in ancient India.

悬空寺

> 第二天，小龙、大卫和林教授来到山西大同悬空寺。大家站在观景台上看着半山腰的寺庙。

大　　卫：好险啊，感觉那个寺庙要掉下来似的。

小　　龙：不会的，你看好多柱子立在寺庙下面呢。

林教授：实际上起关键作用的不是这些立柱，而是那些在寺庙下面横着的飞梁。到跟前可以看得更清楚些。

大　　卫：悬空寺和应县木塔哪个更古老？

林教授：从始建年代看，悬空寺更早些。它始建于公元491年，距今已1500多年了，是国内现存最早、保存最完好的高空木结构摩崖建筑。

大　　卫：名字听起来好复杂。听说《时代》周刊把它评为世界上最不稳定的十大建筑之一。

林教授：是啊，全世界都在为它担着心呢。险峻是它名声在

悬空寺　The Hanging Temple

外的主要原因。这里规模不大,但在建筑技艺方面巧夺天工。

大　卫:你们看寺下面那个大石头。上面好像是"壮观"两个字,可是那个"壮"字右边怎么多了一个点呀?

林教授:这是唐朝大诗人李白题的"壮观"。

大　卫:就是那个写"举头望明月,低头思故乡"的李白吗?

林教授:是的。关于带点儿的"壮"字,有个说法是,当年李白来到悬空寺,感叹它奇险壮观,当场写下了"壮观"两个大字。为了表达自己惊讶至极的心情,就在"壮"字右边加了个点儿。

大　卫：就是表示太壮观了，对吧？

林教授：哈哈，这个解释好。

小　龙：这屋顶好像是琉璃瓦歇山顶。我们在这里还能看见中间那个大殿屋檐下的斗拱呢。看来这个寺庙规格挺高的。

林教授：当初是北魏皇家建造的，应该算是皇家寺院吧。

> 大家来到窄窄的入口处，抬头看见4根柱子支撑着一个亭子，还有一个柱子支撑着一侧的檐角。

大　卫：你们看，这些柱子也太细了吧，居然撑了1500年。

林教授：这种支柱不是一般的木头，是材质坚硬的铁杉木，而且它们不是悬空寺的主要支撑。那些亭子下面的横梁才是主要的承重结构，它们叫半插飞梁。为了防腐、防虫蛀，这些梁柱都经过桐油浸泡。

大　卫：原来是这样啊。

> 大家沿着陡峭的石阶进了寺庙。石阶的左边是山体，多数地方只能容一人前行。他们参观了一圈，回到山下，抬头清楚地看到整个悬空寺下方飞梁伸出山体外的部分。

小　龙：为什么要把寺庙建在悬崖峭壁上呢？

林教授：当初这是个道观，主要是利用山体原有的一处天然凹槽建成。把道观建在半山腰上，就是为了让道士们远离尘世，做到"不闻鸡鸣犬吠之声"。

大　卫：工匠们好聪明呀。

林教授：聪明的方法还在后面呢。工匠们依靠绳索从高处悬下，扩大凹槽，凿出一个可以工作的平台。再在平台上凿出内大外小的巨大石孔。然后，他们用50厘米见方的铁杉木做成方梁，浸过桐油后，把一头带楔子的梁砸入石孔中，楔子便会将木材撑开。

小　龙：像今天膨胀螺栓的原理。

林教授：的确是这样。这些木梁超过三分之二的长度深入山体，加上岩石平台的支撑，每根梁可以承受数吨的重量，所以不用担心安全问题。

大　卫：真是很神奇！

林教授：那些细柱子有的起承重作用，有的用来平衡楼阁的高低，有的在横梁承重超过一定限度时发挥支撑作用。再加上用榫卯结构固定和连接，悬空寺就成了一座似虚而实、似危而安的奇特建筑。

小　　龙：林教授，是飞梁、立柱、榫卯这些木构件的多重保障才支撑起了这个悬空寺，对吗？

林教授：没错。我们这一路从故宫到木塔，再到悬空寺实地考察，是希望你们全面了解中国传统木结构建筑的精妙之处。

大　　卫：太了不起了，太壮观了。我看得在"壮"字右边加上两个点儿。

小　　龙：大卫，现在你的中文水平大涨，说话很有幽默感。

大　　卫：哈哈，这是大诗人李白给我的灵感。

小　　龙：林教授，这次木结构建筑之旅，让我们了解了这种建筑的特别之处，学到了很多有关中国古典建筑的知识。谢谢您。

大　　卫：谢谢林教授，您辛苦了。

The Hanging Temple

> Xiaolong, David, and Prof. Lin went to the Hanging Temple the next day. Standing on the observation platform, they saw the hillside temple.

David: The temple is located on the steep cliff! It seems that it could fall at any time.

Xiaolong: I don't think so. Look, many columns are supporting it.

Prof. Lin: Well, in fact, the columns are not the key. It's those flying beams under the temple that support it. We can see the structure more clearly when we get closer.

David: Which is more ancient, the Hanging Temple or Yingxian Wooden Pagoda?

Prof. Lin: The Hanging Temple. It was built in 491, more than

1,500 years ago. It is the most ancient and best-preserved timber-frame cliff architecture in China.

David: Wow. Its name sounds complicated. By the way, I heard that it is listed as "one of the ten most dangerous buildings in the world" by *TIME* magazine.

Prof. Lin: You're right. The whole world is worrying about it. This temple is famous for its precipitous location. Its architecture craftsmanship is ingenious though it's relatively small in scale.

David: Look at that big rock under the temple. I think it's inscribed with two Chinese characters — *zhuang* (壯) and *guan* (观), collectively meaning spectacular. But there is an extra dot on the right of the Chinese character *zhuang*.

Prof. Lin: Actually, it was inscribed by Li Bai, a great poet in the Tang Dynasty.

David: Li Bai? Is he the poet who wrote the verse, "Looking up, I find the moon bright. Bowing, in homesickness I'm drowned"?

Prof. Lin: Yes. As for the Chinese character *zhuang* with one more dot, legend has it that when Li Bai was visiting this temple, he marveled at its dangerous position and magnificence, and thus wrote those two Chinese characters on the spot. To express his extreme surprise, he added an extra dot to the character *zhuang*.

David: So, that dot is used to stress his surprise, right?

Prof. Lin: That's a good explanation.

Xiaolong: The roof of this temple seems to be a gable and hip roof of glazed tiles. I can see interlocking brackets under the eaves of the main hall. It seems that this temple holds a high rank.

Prof. Lin: It was built by the royal family of the Northern Wei Dynasty, so it's a royal temple.

> Standing at the narrow entrance, they looked up and saw a pavilion propped up by four columns and a corner of the eaves supported by another column.

David: Look! How thin these columns are! Incredible that they have propped up this building for 1,500 years.

Prof. Lin: These columns were not made of ordinary wood, but of rock-hard hemlock timber. Yet, they're not the main supports. Those beams under the pavilion bear the weight of this temple. Professionally, they're called semi-inserted flying beams. They're also immune to corrosion and insects because they were soaked in tung oil beforehand.

David: I see.

> They went along the steep stone steps into the temple. On the left of the stone steps is the mountain, and most steps are wide enough only for one person to pass. After the sightseeing, back at the bottom of the mountain, they looked up and saw clearly the flying beams that extend beyond the mountain under the temple.

Xiaolong: Why did they build it on a cliff?

Prof. Lin: They were asked to build an unearthly Taoist temple where "no sound of chickens or dogs could be heard". The craftsmen took advantage of a natural groove and built the temple halfway up the mountain.

David: How talented they were!

Prof. Lin: Actually, there is something more to show their talents. Suspended from the summit by ropes, the craftsmen used chisels to enlarge the groove to make a platform. Then, standing on the platform, they hollowed out big holes in the wall of the mountain, with the bottom wider than the entrance. Next, they made hemlock timber measuring 50 centimetres in diametre into square beams. After soaking the timber in tung oil, they struck the wedge end of the beams into the stone holes, and the beams would be stuck in the holes firmly.

Xiaolong: This reminds me of the expansion bolts today.

Prof. Lin: Exactly! More than two thirds of these wood beams were penetrated deep into the mountain and supported by the rock platform at the same time. Each beam could bear tons of weight, so no worries about safety.

David: Stunning!

Prof. Lin: Some of these thin columns bear the weight; others balance the heights of the pavilions; and still other columns are for precautions against overweight situations. With the reinforcement of the mortise-and-tenon joints, this temple becomes a unique solid and safe building although looking flimsy and dangerous.

Xiaolong: Prof. Lin, the flying beams, columns, and mortise-and-tenon joints are multiple factors that support this temple, right?

Prof. Lin: Yes. Now we have visited the Forbidden City, Yingxian Wooden Pagoda, and the Hanging Temple, I hope you have a comprehensive understanding of ancient Chinese timber-frame architecture.

David: All these buildings are so amazing and spectacular! I would add two extra dots to the Chinese character *zhuang*.

Xiaolong: That's a good one, David! You are so humorous. Your Chinese has improved greatly.

David: I was inspired by the poet Li Bai.

Xiaolong: Prof. Lin, we have seen the uniqueness of timber-frame architecture and gained much knowledge in this tour. Thank you.

David: Thank you very much, Prof. Lin.

结束语

中国传统木结构建筑营造技艺即木作，是中国建筑传统"八大作"的重要构成，体现了中国人认识自然、利用自然的生活理念，是中华文化中极具代表性的一部分。其工艺的各个环节渗透着中国人的聪明才智，反映了中国人的审美情趣和传统哲学思想。

Summary

Traditional Chinese timber-frame architecture craftsmanship, also known as woodworking, is a critical one of the "Eight Great Feats" in traditional Chinese architecture. It typically represents the Chinese culture and reflects the Chinese concepts of understanding and harnessing the power of nature. This craftsmanship mirrors the ingenuity, aesthetics and philosophies of the Chinese people.

中国历史纪年简表
A Brief Chronology of Chinese History

夏	Xia Dynasty			c. 2070—1600 B.C.
商	Shang Dynasty			1600—1046 B.C.
周	Zhou Dynasty	西周	Western Zhou Dynasty	1046—771 B.C.
		东周	Eastern Zhou Dynasty	770—256 B.C.
		春秋	Spring and Autumn Period	770—476 B.C.
		战国	Warring States Period	475—221 B.C.
秦	Qin Dynasty			221—206 B.C.
汉	Han Dynasty	西汉	Western Han Dynasty	206 B.C.—25
		东汉	Eastern Han Dynasty	25—220
三国	Three Kingdoms			220—280
西晋	Western Jin Dynasty			265—317
东晋	Eastern Jin Dynasty			317—420
南北朝	Northern and Southern Dynasties	南朝	Southern Dynasties	420—589
		北朝	Northern Dynasties	386—581
隋	Sui Dynasty			581—618
唐	Tang Dynasty			618—907
五代	Five Dynasties			907—960
宋	Song Dynasty			960—1279
辽	Liao Dynasty			907—1125
金	Jin Dynasty			1115—1234
元	Yuan Dynasty			1206—1368
明	Ming Dynasty			1368—1644
清	Qing Dynasty			1616—1911
中华民国	Republic of China			1912—1949
中华人民共和国	People's Republic of China			1949—

图书在版编目（CIP）数据

中国传统木结构建筑营造技艺：汉英对照/郭启新，崔红叶主编．－－南京：南京大学出版社，2024.8
（中国世界级非遗文化悦读系列/魏向清，刘润泽主编．寻语识遗）
ISBN 978-7-305-26377-4

Ⅰ．①中… Ⅱ．①郭…②崔… Ⅲ．①木结构－建筑艺术－介绍－中国－汉、英 Ⅳ．① TU-881.2

中国版本图书馆 CIP 数据核字（2022）第 233927 号

出版发行	南京大学出版社
社　　址	南京市汉口路 22 号　　　邮　编　210093
丛 书 名	中国世界级非遗文化悦读系列·寻语拾遗
丛书主编	魏向清　刘润泽
书　　名	**中国传统木结构建筑营造技艺：汉英对照**
	ZHONGGUO CHUANTONG MUJIEGOU JIANZHU YINGZAO JIYI: HANYING DUIZHAO
主　　编	郭启新　崔红叶
责任编辑	张淑文　　　编辑热线　（025）83592401
照　　排	南京新华丰制版有限公司
印　　刷	南京凯德印刷有限公司
开　　本	880mm×1230mm　1/32 开　印张 6　字数 124 千
版　　次	2024 年 8 月第 1 版　2024 年 8 月第 1 次印刷
ISBN 978-7-305-26377-4	
定　　价	69.00 元

网址：http://www.njupco.com
官方微博：http://weibo.com/njupco
官方微信号：njupress
销售咨询热线：（025）83594756

* 版权所有，侵权必究
* 凡购买南大版图书，如有印装质量问题，请与所购图书销售部门联系调换